Randomness and Elements of Decision Theory Applied to Signals

Monica Borda · Romulus Terebes · Raul Malutan ·
Ioana Ilea · Mihaela Cislariu · Andreia Miclea ·
Stefania Barburiceanu

Randomness and Elements of Decision Theory Applied to Signals

Monica Borda
Technical University of Cluj Napoca
Cluj-Napoca, Romania

Raul Malutan
Technical University of Cluj-Napoca
Cluj-Napoca, Romania

Mihaela Cislariu
Technical University of Cluj-Napoca
Cluj-Napoca, Romania

Stefania Barburiceanu
Technical University of Cluj-Napoca
Cluj-Napoca, Romania

Romulus Terebes
Technical University of Cluj-Napoca
Cluj-Napoca, Romania

Ioana Ilea
Technical University of Cluj-Napoca
Cluj-Napoca, Romania

Andreia Miclea
Technical University of Cluj-Napoca
Cluj-Napoca, Romania

ISBN 978-3-030-90316-9   ISBN 978-3-030-90314-5   (eBook)
https://doi.org/10.1007/978-3-030-90314-5

This Springer imprint is published by the registered company Springer Nature Switzerland AG
The registered company address is: Gewerbestrasse 11, 6330 Cham, Switzerland

Preface

The work, organized in 15 chapters and 3 Appendices, presents the basics of random variables/processes and decision theory, all these massively illustrated with examples, mainly from telecommunications and image processing.

The addressability of the book is principally to students in electronics, communications and computer sciences branches, but we think it is very useful to Ph.D. students dealing with applications requiring randomness in many other areas.

The whole material has an educational presentation, starting with theoretical backgrounds, followed by simple examples and MATLAB implementation and ending with tasks given to the reader as homework. A short bibliography is available for each chapter.

The difficulty of the material is put in a progressive way, starting with simple examples, and continuing with applications taken from signal and image processing.

Chapter 1, *Random variables*, introduces random variables and the study of their characteristic functions and values, from both theoretical and practical (MATLAB simulations) points of view.

Chapter 2, *Probability distributions*, aims to study the probability distributions and the central limit theorem, both theoretically and practically using MATLAB.

Chapter 3, *Joint random variables*, describes joint random variables and their characteristic functions and values. These elements are theoretically introduced and a MATLAB implementation is available.

Chapter 4, Random processes, describes the random processes theoretically but also practically using MATLAB.

Chapter 5, Pseudo-noise sequences. The purpose of this work is the study of the properties and applications of the Pseudo-Noise Sequences (PNS). MATLAB implementation of a PNS generator is also proposed in the chapter.

Chapter 6, Markov systems, shows several problems and exercises to illustrate the theory of memory sources and Markov chains theory.

Chapter 7, Noise in telecommunication systems. The purpose of this chapter is the study of the noise that appears in communication systems.

Chapter 8, Decision systems in noisy transmission channels, aims the study of decision systems in noisy transmissions and the implementation of an application with a graphical user interface in the LabVIEW environment.

Chapter 9, Comparison of ICA algorithms for oligonucleotide microarray data, presents a study and comparison of Independent Component Analysis algorithms on oligonucleotide microarray data, with application in MATLAB.

Chapter 10, Image classification based on statistical modeling of textural information, introduces an application of statistics, for image processing. More precisely, a classification algorithm based on the statistical modeling of textural information is presented.

Chapter 11, Histogram equalization, aims to study the histogram of an image and the process of histogram equalization of an image.

Chapter 12, PCM and DPCM—applications in image processing, describes pulse code modulation (PCM) and differential pulse code modulation (differential PCM or DPCM) with applications in image processing.

Chapter 13, Nearest neighbor (NN) and k-nearest neighbor (kNN) supervised classification algorithms, presents the theoretical foundations associated with the supervised classification algorithms of NN (*Nearest Neighbor*) and kNN (*k-Nearest Neighbors*) type and exemplifies their use in data classification applications using MATLAB, operating on simplified representations of forms and patterns in terms of feature vectors having the most representative characteristics (descriptors).

Chapter 14, Texture feature extraction and classification using the Local Binary Patterns operator, shows the theoretical fundamentals regarding texture feature extraction using the Local Binary Patterns operator as well as its use in a MATLAB application that performs image classification.

Chapter 15, Supervised deep learning classification algorithms, aims to present the theoretical foundations associated with supervised classification networks of deep learning (CNNs), as well as to exemplify their use in data classification applications in MATLAB.

The examples were selected to be as simple as possible, but pointing out the essential aspects of the processing. Some of them are classical, others are taken from the literature, but the majority are original.

The understanding of the phenomenon, of the aim of processing, in its generality, not necessarily linked to a specific application, the development of the "technical good sense" is the logic thread guiding the whole work.

Cluj-Napoca, Romania   Monica Borda
                                                                            Romulus Terebes
                                                                               Raul Malutan
                                                                                  Ioana Ilea
                                                                            Mihaela Cislariu
                                                                              Andreia Miclea
                                                                       Stefania Barburiceanu

Contents

List of Figures

Chapter 1
Random Variables

1.1 Definitions

Definition: *A signal, a phenomenon, or an experiment is random* if its development is, at least partially, ruled by probabilistic laws, meaning that the result (also called outcome, sample, or event) of such experiment is confirmed only by accomplishment.

Definition: *A random variable* (r.v.) X is a function defined on the sample space S (the set of all possible distinct outcomes, samples, or events $s_i \in S$) and taking real values:

$$X : S \to \mathbb{R}$$

$$X(s_i) = x_i$$

where: $x_i \in \mathbb{R}$ is the value of the random variable $X(s_i)$.

The random variables are denoted by upper case letters (e.g. X), while their values are denoted by the corresponding lower-case letters (e.g. x).

Remark

M. Borda et al., *Randomness and Elements of Decision Theory Applied to Signals*,
https://doi.org/10.1007/978-3-030-90314-5_1

1.2 Classification

1.2.1 Discrete Random Variables

Definition: *A random variable X is discrete* if the number of possible values is finite (for instance $x_i \in \{0, 1\}$), or countably infinite (such as $x_i \in \{0, 1, 2, \cdots\}$, $x_i \in \mathbb{Z}$) [1].

Tossing a die, flipping a coin, or symbol generation by a binary source are some examples of events that can be modeled by discrete random variables [2].

Example

1.2.2 Continuous Random Variables

Definition: *A random variable X is continuous* if it can take an infinite number of values from an interval of real numbers [1].

The noise voltage in a resistor or the lifetime of electronic devices are some examples of phenomena modeled by continuous random variables.

Example

1.3 Characteristic Functions

1.3.1 Probability Distribution Function

Abbreviation: PDF, Or cdf (*Cumulative Distribution Function*)

Definition: Let X be a random variable. The *probability distribution function* of X is denoted by $F_X(x)$ and it represents the function $F_X : \mathbb{R} \to [0, 1]$, defined as:

$$F_X(x) = P\{X \le x\}, \forall x \in \mathbb{R}$$

Properties:

- $F_X(x) \ge 0, \forall x \in \mathbb{R}$;
- $F_X(-\infty) = 0$;
- $F_X(\infty) = 1$;
- $F_X(x)$ is not a decreasing function : $x_1 < x_2 \Rightarrow F_X(x_1) \le F_X(x_2)$.

If X is a discrete random variable, then a probability p_i is associated to each sample s_i: $p_i = p(s_i) = P\{X(s_i) = x_i\} = p(x_i)$. In this case, the

probability mass function (PMF) is defined as: $X : \begin{pmatrix} x_i \\ p_i \end{pmatrix}, \sum_i p_i = 1$

Remark

1.3.2   Probability Density Function

Abbreviation: pdf.

Definition: Let X be a random variable. The *probability density function* of X is denoted by $f_X(x)$ and it represents the derivative of the probability distribution function:

$$f_X(x) = \frac{d F_X(x)}{dx}$$

Properties:

- $f_X(x) \ge 0, \forall x \in \mathbb{R}$;
- $F_X(x) = \int\limits_{-\infty}^{x} f_X(t)dt$

- $\int\limits_{-\infty}^{\infty} f_X(x)dx = F_X(\infty) - F_X(-\infty) = 1$;

- $\int\limits_{x_1}^{x_2} f_X(x)dx = \int\limits_{-\infty}^{x_2} f_X(x)dx - \int\limits_{-\infty}^{x_1} f_X(x)dx = F_X(x_2) - F_X(x_1) =$

$$= P(x_1 < X \le x_2)$$

The probability density function can be defined only if the probability distribution function is continuous, and it has a derivative

Remark

1.4   Characteristic Values

1.4.1   nth Order Average Operator

Definition: Let X be a random variable. The nth *order average operator* is denoted by $E\{X^n\}$ and its formula depends on the variable's type:

- if X is a discrete random variable:

$$E\{X^n\} = \sum_i x_i^n p(x_i)$$

- if X is a continuous random variable:

$$E\{X^n\} = \int_{-\infty}^{\infty} x^n f_X(x)dx$$

The most used average operators:

- If $n = 1$, the *expected value*, or the *expectation*, or the *mean* of the random variable is obtained, and it is denoted by $\mu_X = E\{X\}$.

Definition: The *expected value* represents the mean of all possible values taken by the random variable, weighted by their probabilities.
  Properties:

- $E\{a\} = a$, if $a \in \mathbb{R}$ is a constant;
- $E\{ax\} = aE\{X\}$, if $a \in \mathbb{R}$ is a constant;
- $E\{ax + b\} = aE\{X\} + b$, if $a, b \in \mathbb{R}$ are constants;
- The operator can be applied also to functions depending on X. Let consider the function $g(X)$. In this case, the expectation is given by:

$$E\{g(X)\} = \begin{cases} \sum_i g(x_i)p(x_i), & X \text{ discrete r.v.} \\ \int\limits_{-\infty}^{\infty} g(x)f_X(x)dx, & X \text{ continuous r.v.} \end{cases}$$

The mean can be interpreted as the DC component of a signal.

Remark

- If $n = 2$, the second order average operator is obtained, denoted by $E\{X^2\}$.

The significance of this operator is that of the power of a signal.

Remark

1.4.2 nth Order Centered Moment

Definition: Let X be a random variable. The nth *order centered moment* of the random variable is denoted by $E\{(X - E\{X\})^n\}$ and its formula depends on the variable's type:

- if X is a discrete random variable:

$$E\{(X - E\{X\})^n\} = \sum_i (x_i - E\{X\})^n p(x_i);$$

- if X is a continuous random variable:

$$E\{(X - E\{X\})^n\} = \int\limits_{-\infty}^{\infty} (x - E\{X\})^n f_X(x)dx,$$

where $E\{X\}$ is the expectation.

- $X - E\{X\}$ represents the amplitude of the random variable's oscillation around the mean.
- $E\{X - E\{X\}\} = 0$.

Remark

The most used centered moment: If $n = 2$, the second order centered moment, also called *variance*, or *dispersion* is obtained, and it is denoted by $\sigma_X^2 = E\{(X - E\{X\})^2\}$.

- The significance of this operator is that of the variable's power of the oscillation around the mean.
- A small value of the dispersion implies that all the variable's values are close to their mean $E\{X\}$.
- The dispersion is not a linear operator.
- The *standard deviation* is defined as: $\sigma = \sqrt{\sigma_X^2}$

Remark

Properties:

- $\sigma_X^2\{a\} = 0$, if $a \in \mathbb{R}$ is a constant;
- $\sigma_X^2\{ax\} = a^2\sigma_X^2(x)$, if $a \in \mathbb{R}$ is a constant;
- $\sigma_X^2\{ax + b\} = a^2\sigma_X^2(x)$, if $a, b \in \mathbb{R}$ are constants.

1.5 Other Statistical Means

When studying samples, some other statistical means can be used, including the arithmetic mean, the geometric mean, the harmonic mean, the median, the trimmed average, or the Fréchet mean.

Let X be a random variable and $x = \{x_1, x_2, \ldots, x_N\}$ a sample of size N. In this case, the following measures can be defined:

(a) **Arithmetic mean:**

$$\overline{x}_{MA} = \frac{1}{N}\sum_{i=1}^{N} x_i.$$

Remark

- The arithmetic mean can be viewed as the sample's *center of mass* (or *centroid*), and it is obtained by minimizing the Euclidean distance between the elements $x_1, x_2, \ldots, x_N$.

(b) Geometric mean:

$$\bar{x}_{MG} = \sqrt[N]{\prod_{i=1}^{N} x_i}.$$

Example

Let consider $x = \{1, 9, 7, 3, 5\}$. Knowing that $N = 5$, the geometric mean is: $\bar{x}_{MG} = \sqrt[5]{1 \times 9 \times 7 \times 3 \times 5} = 3.93$.

(c) Harmonic mean:

$$\bar{x}_{MH} = N \left(\sum_{i=1}^{N} \frac{1}{x_i} \right)^{-1}.$$

Example

Let consider $x = \{1, 9, 7, 3, 5\}$. Knowing that $N = 5$, the harmonic mean is:

$$\bar{x}_{MH} = 5 \left(\tfrac{1}{1} + \tfrac{1}{9} + \tfrac{1}{7} + \tfrac{1}{3} + \tfrac{1}{5} \right)^{-1} = 2.79.$$

Remark

- Between the previously defined means (arithmetic, geometric and harmonic) the following relation can be established:
  $$\bar{x}_{MA} \geq \bar{x}_{MG} \geq \bar{x}_{MH}.$$

(d) **Median:** This value is obtained by sorting the elements $x_1, x_2, \ldots, x_N$ in ascending order and by taking the central value that divides the initial set into two subsets. Both the median and the arithmetic mean are centroid computation methods.

Compared to the arithmetic mean, the median is less influenced by the *outlier values* (values strongly different from the majority), giving a much better estimate of the central value.

Let consider $x = \{1, 9, 7, 3, 5\}$. To compute the median, the vector is first sorted in ascending order, giving the set $\{1, 3, 5, 7, 9\}$. Next, the median is obtained by choosing the central value, which for this example is 5.

Example

- If the sample x contains an even number of elements, then the median is computed as the arithmetic mean of the two central values.

Remark

(e) **Trimmed average:** This method is used when the data set $\{x_1, x_2, \ldots, x_N\}$ contains outlier values. In order to reduce their impact on the centroid estimation, the trimmed average can be employed. This approach deals with outliers by eliminating them from the dataset. Next, it computes the arithmetic mean or the median of the remaining elements.

(f) **Fréchet mean (or Karcher mean):** Represents a method to compute the center of mass, when the samples are defined on a surface.

The covariance matrices can be modeled as realizations of a random variable defined on a surface. In this case, the center of mass can be computed by using the Fréchet mean.

Example

- The arithmetic mean, the geometric mean and the harmonic mean can be viewed as Fréchet means for real-valued data.

Remark

1.6 Applications and Matlab Examples

1.6.1 Computation of Characteristic Functions and Values

Exercise: Let consider the experiment consisting in tossing a fair die. The result is observed, and the number of obtained dots is reported. Compute the probability mass function and the probability distribution function characterizing the random variable that models the experiment. What is the expectation of this variable? What is its dispersion?

Solution:

Let X be the discrete random variable modeling the experiment. The sample space S contains the six dice faces $S = \{f_1, \ldots, f_6\}$. The values taken by the variable X are given by the number of dots on each face: $x = \{1, \ldots, 6\}$. Supposing that a fair die is used, all the possible outcomes have the same probability: $p_i = \frac{1}{6}, \forall i = \overline{1, 6}$ [3].

Therefore, the probability mass function describing the random variable X is:

$$X : \begin{pmatrix} 1 & 2 & 3 & 4 & 5 & 6 \\ \frac{1}{6} & \frac{1}{6} & \frac{1}{6} & \frac{1}{6} & \frac{1}{6} & \frac{1}{6} \end{pmatrix}$$

The probability distribution function must be computed for each interval:

$$\forall x \in (-\infty, 1) : F_X(x) = P\{X \leq x\} = 0;$$

$$\forall x \in [1, 2) : F_X(x) = P\{X \leq x\} = \sum_{x_i \leq x} p_i = \frac{1}{6};$$

$$\forall x \in [2, 3) : F_X(x) = P\{X \leq x\} = \sum_{x_i \leq x} p_i = \frac{1}{6} + \frac{1}{6} = \frac{2}{6};$$

$$\forall x \in [3, 4) : F_X(x) = P\{X \leq x\} = \sum_{x_i \leq x} p_i = \frac{1}{6} + \frac{1}{6} + \frac{1}{6} = \frac{3}{6};$$

$$\forall x \in [4, 5) : F_X(x) = P\{X \leq x\} = \sum_{x_i \leq x} p_i = \frac{1}{6} + \frac{1}{6} + \frac{1}{6} + \frac{1}{6} = \frac{4}{6};$$

$$\forall x \in [5, 6) : F_X(x) = P\{X \le x\} = \sum_{x_i \le x} p_i = \frac{1}{6} + \frac{1}{6} + \frac{1}{6} + \frac{1}{6} + \frac{1}{6} = \frac{5}{6};$$

$$\forall x \in [6, \infty) : F_X(x) = P\{X \le x\} = \sum_{x_i \le x} p_i = \frac{1}{6} + \frac{1}{6} + \frac{1}{6} + \frac{1}{6} + \frac{1}{6} + \frac{1}{6} = 1.$$

The expected value:

$$\mu_X = E\{X\} = \sum_{i=1}^{6} x_i p_i = 1 \times \frac{1}{6} + 2 \times \frac{1}{6} + 3 \times \frac{1}{6}$$

$$+ 4 \times \frac{1}{6} + 5 \times \frac{1}{6} + 6 \times \frac{1}{6} = 3.5$$

The dispersion:

$$\sigma_X^2 = E\{(X - E\{X\})^2\} = \sum_{i=1}^{6} (x_i - \mu_X)^2 p_i =$$

$$= (1 - 3.5)^2 \times \frac{1}{6} + (2 - 3.5)^2 \times \frac{1}{6} + (3 - 3.5)^2 \times \frac{1}{6} +$$

$$+ (4 - 3.5)^2 \times \frac{1}{6} + (5 - 3.5)^2 \times \frac{1}{6} + (6 - 3.5)^2 \times \frac{1}{6} = 2.93.$$

Matlab code:

Initialization of the random variable's values and their corresponding probabilities:

```
% Values taken by the random variable X:
x_i = [1, 2, 3, 4, 5, 6];
% Their probabilities:
p_xi = [1/6, 1/6, 1/6, 1/6, 1/6, 1/6];
```

Drawing the probability mass function implies the graphical representation of the probabilities p_i as functions of the random variable's values x_i. In order to represent discrete values, Matlab proposes the *stem* function:

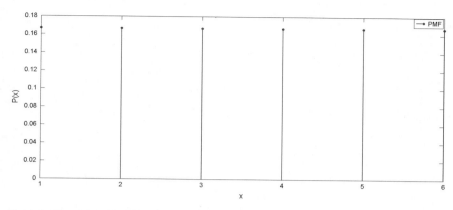

Fig. 1.1 Plot of the probability mass function

```
% PMF
% Create new figure
figure;
% Plot the probabilities
stem(x_i, p_xi, 'o', 'MarkerFaceColor', 'blue', 'Lin-
eWidth', 1.5)
% Set X-axis label
xlabel('x')
% Set values on X axis
set(gca,'XTick',x_i)
% Set Y-axis label
ylabel('P(x)')
% Add the legend of the plot
legend('PMF')
```

In Fig. 1.1 it can be noticed that all the six values taken by the random variable *X* have the same probability equaling 1/6.

The probability distribution function can be obtained by using the *cumsum* Matlab function, which gives the cumulative sum of a vector:

```
% PDF
% Compute the probability distribution function
F_xi = cumsum(p_xi);
```

Knowing that the PDF can take values between 0 and 1, the limits of this interval must be added on the plot:

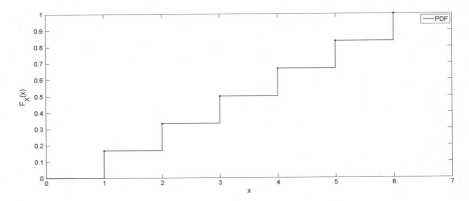

Fig. 1.2 Plot of the probability distribution function

```
% Add two new values on the plot
xi_grafic = [min(x_i)-1, x_i, max(x_i)+1];
F_xi_grafic = [0, F_xi, 1];

% Create new figure
figure;
% Plot the PDF
stairs(xi_grafic, F_xi_grafic, 'LineWidth', 1.5)
hold on
% Mark the included limit points
plot(x_i, F_xi, 'o', 'MarkerFaceColor', 'blue',
'MarkerSize', 5)
% Set X-axis label
xlabel('x')
% Set Y-axis label
ylabel('F_X(x)')
% Add the legend of the plot
legend('PDF')
```

The discrete aspect of the random variable can be noticed in Fig. 1.2, where the PDF is a staircase function or a step function.

1.6.2 Influence of Outlier Values on the Data's Centroid

Arithmetic mean versus median: The arithmetic mean and the median are some of the most used methods for the centroid estimation. In the following, their behavior will be analyzed when the dataset contains outlier values.

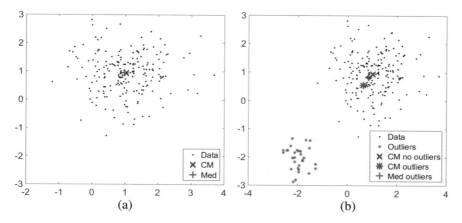

Fig. 1.3 Behavior of the arithmetic mean and median **a** for an outlier free dataset and **b** in the presence of outliers

If there are no outliers, the mean (CM) and the median (Med) perform almost identically, giving very close results, as it can be seen in Fig. 1.3a. Further on, if outliers are added, the centroid obtained by computing the arithmetic mean is attracted by them. On the other hand, the median gives a centroid that remains much closer to the initial value, as illustrated in Fig. 1.3b.

Trimmed average: This is another method used to compute the data's centroid by eliminating the outliers from the dataset. For this approach, the proportion α of ignored data is fixed. The observations are eliminated according to a criterion. For instance, the arithmetic mean, or the median of the initial set is computed and $\alpha\%$ of the farthest observations are discarded. The final centroid is given in the end by the mean or the median of the newly obtained set.

The arithmetic mean (CM) and the trimmed averages (Trim) for $\alpha \in$ $\{0\%, 2\%, 5\%, 10\%\}$ are illustrated in Fig. 1.4. All these results are compared to those obtained for the arithmetic mean computed on the initial dataset in Fig. 1.4a. It can be noticed that by increasing the percentage of discarded outliers, the trimmed average becomes closer to the centroid of the initial data.

The main disadvantage of this approach is the fact that by eliminating some observations, the number of remaining samples may be too small for reliable statistic results. In addition, by eliminating some outliers, new samples can become aberrant values.

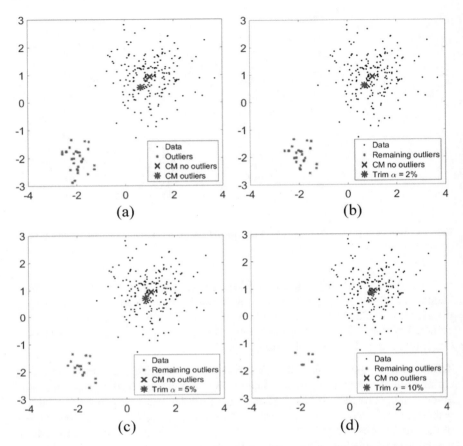

Fig. 1.4 Behavior of the trimmed average for different percentages of discarded outliers **a** $= 0\%$, **b** $\alpha = 2\%$, **c** $\alpha = 5\%$ and **d** $\alpha = 10\%$

1.7 Tasks

Make a document containing the solutions for the following exercises:

1. For the exercise in Sect. 1.6.1, compare the Matlab code with the mathematical solution.
2. Let consider the experiment consisting in observing the parity check bit of a word (8 bits) in the computer memory. This bit can take two values: 0 with probability q, and 1 with probability $1 - q$. Draw the probability mass function and the probability distribution function of the random variable modeling the experiment [4]. Implement the Matlab code.
3. Implement the Matlab code for computing the arithmetic mean, the geometric mean, the harmonic mean, the median and the trimmed average.

1.8 Conclusions

This chapter studies the random variables and their characteristic functions and values. It contains both theoretical aspects (definitions, properties, etc.) and practical elements concerning the implementation in Matlab. In addition, the influence of outliers on the statistical means is illustrated.

References

1. Lee JK (2010) Statistical bioninformatics. John Wiley & Sons Inc., For Biomedical and Life Science Researchers
2. Mitrea A (2006) Variabile şi semnale aleatoare. Press, U.T
3. Ciuc M, Vertan C (2005) Prelucrarea Statistică a Semnalelor. MatrixRom
4. Stark H, Woods JW (2002) Probability and random processes with applications to signal processing. Prentice-Hall

Chapter 2
Probability Distributions

2.1 Discrete Distributions

In the first part, we will present some examples of distributions that can be used with discrete random variables.

2.1.1 Bernoulli Distribution

The Bernoulli distribution models experiments that have two possible results, generally noted with 1, for the successful event, and 0 for the opposite event. The parameter of this distribution is represented by the probability p of the successful event [1].

Definition: A discrete random variable X has a *Bernoulli distribution* of parameter p, if it can take values 0 and 1 with probabilities $1 - p$ and p.

Probability Mass Function:

$$X : \begin{pmatrix} 0 & 1 \\ 1 - p & p \end{pmatrix}.$$

Mean value: $\mu = p$.

Dispersion: $\sigma^2 = (1 - p)p$.

 Bernoulli distribution can model the generation of a symbol by a binary source (for instance, the generation of a 0 bit with the probability equaling p).

Example

2.1.2   Binomial Distribution

Let n independent events, each having a probability of success p. The probability of having k successful events is determined by binomial distribution. In other words, the binomial distribution models the number of successes from n independent Bernoulli events having the parameter p [2].

Definition: A discrete random variable X has a *binomial distribution* of parameters $n \in \mathbb{N}$ and $p \in [0, 1]$, if it can take the values $0, 1, 2, \ldots, n$, with probabilities $P\{X = k\} = C_n^k p^k (1 - p)^{n-k}$, where $k = 0, 1, 2, \ldots, n$.

Probability Mass Function:

$$X : \begin{pmatrix} k \\ C_n^k p^k (1 - p)^{n-k} \end{pmatrix}, k = 0, 1, 2, \ldots, n.$$

Mean value: $\mu = np..$

Dispersion: $\sigma^2 = n(1 - p)p..$

- A random variable with binomial distribution with parameters n and p is the sum of n independent random variables with Bernoulli distribution of parameter p.
- The Bernoulli distribution is a particular case of the binomial distribution, obtained for $n = 1$ [2].

Remark

For a binary symmetric channel, in the case of independent errors, the number of erroneous bits from a word of n bits can be modeled by a binomial distribution.

Example

2.1.3   Poisson Distribution

Let $\lambda > 0$ be the mean value of events occurring within a given period of time. The probability of having k events in that interval is determined by the Poisson distribution.

Definition: A discrete random variable X has a *Poisson distribution* of parameter $\lambda > 0$, if it can take any positive integer value, with probabilities $P\{X = k\} = \frac{\lambda^k}{k!}e^{-\lambda}$, where $k = 0, 1, 2, \ldots$.

Probability Mass Function:

$$X : \begin{pmatrix} k \\ \frac{\lambda^k}{k!} e^{-\lambda} \end{pmatrix}, k = 0, 1, 2, \ldots.$$

Mean value: $\mu = \lambda$.

Dispersion: $\sigma^2 = \lambda$.

- Poisson distribution is a limiting case of the binomial distribution obtained when $n \to \infty$, $p \to 0$ and $np \to \lambda$[2].

Remark

The number of calls received in a call center during a day can be modeled by the Poisson distribution.

Example

2.1.4 Geometric Distribution

Let consider a sequence of events, each having the probability of success equaling p. The number of events required until the first successful event is determined by the geometric distribution.

Definition: A discrete random variable X has a *geometric distribution* of parameter $p \in (0, 1]$, if it can take any positive integer values with probabilities $P\{X = k\} = (1 - p)^k p$, where $k = 0, 1, 2, \ldots.$

Probability Mass Function:

$$X : \begin{pmatrix} k \\ (1 - p)^k p \end{pmatrix}, k = 0, 1, 2, \ldots.$$

Mean value: $\mu = \frac{1-p}{p}$.

Dispersion: $\sigma^2 = \frac{1-p}{p^2}$.

- Geometric distribution models the number of independent Bernoulli events that occur until the first successful event.

Remark

On a noisy channel, the number of times that a message is transmitted until the first correct reception can be modeled by a geometric distribution.

Example

2.2   Continuous Distributions

2.2.1   Uniform Distribution

Definition: A continuous random variable X has a *uniform distribution* over the interval $[a, b]$ if its probability density function is:

$$f_X(x) = \begin{cases} \frac{1}{b-a}, & a \leq x \leq b \\ 0, & \text{otherwise} \end{cases}$$

.

Mean value: $\mu = \frac{a+b}{2}$.

Dispersion: $\sigma^2 = \frac{(b-a)^2}{12}$.

For analog-to-digital converters, the quantization error has a uniform distribution.

Example

2.2.2   Normal Distribution (Gaussian)

Definition: A continuous random variable X has a *normal* or *Gaussian distribution* of parameters $\mu \in \mathbb{R}$ and $\sigma > 0$ if its probability density function is:

$$f_X(x) = \frac{1}{\sqrt{2\pi\sigma^2}} e^{-\frac{(x-\mu)^2}{2\sigma^2}}.$$

Mean value: μ.

Dispersion: σ^2.

In a communications system, the channel's noise can be modeled by a Gaussian distribution of mean value $\mu = 0$ and dispersion σ^2.

Example

- A particular case of the normal distribution is the *standard normal distribution*, obtained for $\mu = 0$ and $\sigma = 1$. In this case, the probability density function is:

$$f_X(x) = \frac{1}{\sqrt{2\pi}} e^{-\frac{1}{2}x^2}.$$

Remark

2.2.3 Rayleigh Distribution

Definition: A continuous random variable X has a *Rayleigh distribution* of parameter $b > 0$ if its probability density function is:

$$f_X(x) = \begin{cases} \frac{1}{b-a}, & a \leq x \leq b \\ 0, & \text{otherwise} \end{cases}$$

.

Mean value: $\mu = b\sqrt{\frac{\pi}{2}}$.

Dispersion: $\sigma^2 = \frac{4-\pi}{2}b^2$.

The lifetime of electronic components or fading channels can be modeled using the Rayleigh distribution.

Example

2.3   The Central Limit Theorem

Let $X_1, \ldots, X_n$ be a set of random independent variables and identically distributed (i.i.d.), characterized by the same probability distribution with mean μ and dispersion σ^2. In the case where n is large enough:

- The sum of the random variables $S = X_1 + \cdots + X_n$ is a random variable with a normal distribution, having the mean $\mu_S = n\mu$ and the dispersion $\sigma_S^2 = n\sigma^2$.
- The mean of the random variables $\overline{X} = \frac{X_1 + \cdots + X_n}{n}$ is a random variable with normal distribution, having the mean $\mu_{\overline{X}} = \mu$ and the dispersion $\sigma_{\overline{X}}^2 = \frac{\sigma^2}{n}$ [2].

Remark

The remark according to which "a random variable with binomial distribution with parameters n and p is the sum of n independent random variables with Bernoulli distribution of parameter p" (Sect. 2.1.2) does not contradict the central limit theorem. In this case, as p gets lower values (e.g. $p = 0.01$), the binomial distribution loses its symmetry, and it becomes different from a normal distribution. Thus, for a correct approximation with a normal distribution, the lower the value p, the higher the value of n must be [2, 3].

2.4   Probability Distributions in Matlab

In the *Statistics and Machine Learning Toolbox*, Matlab provides users with predefined functions for data modeling. These include methods for determining the probability distribution function (PDF), the density probability function (pdf), as well as functions for generating random samples with a certain distribution.

In the tables below, there are some examples of Matlab functions that can be used to simulate discrete random variables (Table 2.1) and continuous random variables (Table 2.2).

In Matlab, the probability distribution function is denoted by CDF (*Cumulative Distribution Function*).

Remark

Table 2.1 Examples of Matlab functions for the study of discrete distributions

Name	PDF	pdf	Generate samples
Binomial	`binocdf`	`binopdf`	`binornd`
Geometric	`geocdf`	`geopdf`	`geornd`
Poisson	`poisscdf`	`poisspdf`	`poissrnd`

Table 2.2 Examples of Matlab functions for studying continuous distributions

Name	PDF	pdf	Generate samples
Beta	`betacdf`	`betapdf`	`betarnd`
Exponential	`expcdf`	`exppdf`	`exprnd`
Gamma	`gamcdf`	`gampdf`	`gamrnd`
Normal	`normcdf`	`normpdf`	`normrnd`
Rayleigh	`raylcdf`	`raylpdf`	`raylrnd`
Student	`tcdf`	`tpdf`	`trnd`
Weibull	`wblcdf`	`wblpdf`	`wblrnd`

2.5 Applications and Matlab Examples

2.5.1 Probability Distribution Tool

"*Probability Distribution Tool*" [4] is a Matlab application which, by using a graphical interface, allows the user to visualize the probability distribution function and the probability density function for about 24 distributions. Also, the interface allows changing the characteristic parameters of each distribution and analyzing their impact on the aspect of the two functions.

The application can be launched from the command line by using the command *disttool*:

```
>> disttool
```

The graphical interface is shown in Fig. 2.1. The "**Distribution**" menu allows the selection of the desired distribution, and from the "**Function type**" menu the probability distribution function (CDF) or the probability density function (PDF) can be selected. For example, in Fig. 2.1, the probability density function for the normal distribution is presented. The shape of the graph depends on the two parameters μ and σ, which can be changed.

In "*Probability Distribution Tool*", the probability density function is case-insensitive: PDF (*Probability Density Function*)

Remark

2.5.2 Generating Bernoulli Experiments

A realization of a random variable X having a Bernoulli distribution with parameter p can be simulated in Matlab, as follows:

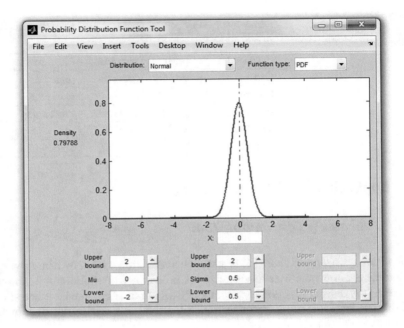

Fig. 2.1 The graphical interface of the application "*Probability Distribution Tool*"

```
clear all

close all

clc


% Definition of parameter p of the Bernoulli distribution:

p = 0.5;

% Generation of a random value from a uniform distribution between 0
and 1:

U = rand(1,1)

% Definition of the random variable X having a Bernoulli distribution
with parameter p, knowing that if U < p, then the experiment was a
success:

X = U < p
```

To simulate n independent realizations of a random variable X with Bernoulli distribution of parameter p, the code becomes:

```
% Number of independent events:

n = 10;

% Generation of n random values from a uniform distribution between 0
and 1:

U = rand(1, n)


% Generation of n realizations of the random variable X having a
Bernoulli distribution with parameter p, knowing that if U < p, then
the experiment was a success:

X = U < p
```

2.5.3   The Connection Between Bernoulli Distribution and Binomial Distribution

In practice, the connection between the two distributions can be illustrated. First, m Bernoulli experiments are generated, each of them having n realizations. This creates a matrix with m lines and n columns. Next, the sum of the elements on each line represents the number of successful events from n Bernoulli events. Thus, m realizations of a binomial distribution of parameters n and p can be obtained. The Matlab code for the simulation is the following:

```
clear all

close all

clc


% Definition of parameter p of the Bernoulli distribution:

p = 0.5;

% Number of independent Bernoulli realizations:

n = 20;

% Number of Bernoulli experiments:

m = 50;

% Generation of m Bernoulli experiments having n realizations:

y = rand(m, n)< p;

% Number of successful events for each of the m experiments:

s = sum(y,2);
```

The normalized histogram of the m sums is an estimation of the probability mass function, and it is calculated as follows:

```
% Computation of the histogram:
[h, x] = hist(s);
% Histogram normalization:
dx = diff(x(1:2));
h = h/sum(h*dx);
% Graphical representation of the histogram:
figure
bar(x,h)
```

For verification, it is possible to overlap the histogram with the probability density function of a random variable with binomial distribution with parameters n and p:

```
hold on
% Computation of the binomial probability density function:
pd = binopdf(0:n,n,p);
% Graphical representation of the probability density function:
plot(0:n,pd, 'r')
% Labels for the axes
xlabel('x')
ylabel('P(x)')
% Legend
legend ('Histogram', 'pdf')
```

Using the previous code, the figures presented in Fig. 2.2 are generated, knowing that $p = 0.05$, $n = \begin{bmatrix} 50 & 100 & 500 \end{bmatrix}$ and $m = 100$. The purpose of the experiment is to show that by increasing the number of independent Bernoulli events, their sum is closer to a binomial distribution.

2.5.4 The Connection Between Bernoulli Distribution and the Geometric Distribution

To illustrate the connection between the two distributions, m Bernoulli experiments are generated, each having n realizations, obtaining a matrix with m lines and n columns. Keeping the notations from Sect. 2.5.3, the matrix is denoted by y, for which $n = 300$ and $m = 100$.

The number of Bernoulli events occurring until the first successful event is determined by calculating the maximum on each line in the matrix. Knowing that each

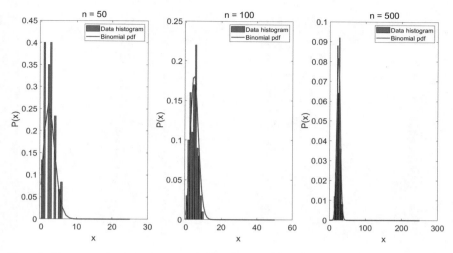

Fig. 2.2 The connection between Bernoulli distribution and binomial distribution

line contains only two possible elements, 0 and 1, the max function will return the maximum value (that is 1) and the position for its first appearance:

```
% Number of realizations until the first success:
[val,index]= max(y,[],2);
```

The normalized histogram of the previously obtained positions (corresponding to the appearances of the maximal value) approximates the probability mass distribution function and it is calculated as follows:

```
% Computation of the normalized histogram
[h, x] = hist(index-1);
dx = diff(x(1:2));
h = h/sum(h*dx);
% Graphical representation of the histogram
figure
bar(x,h)
```

The histogram can be overlaid with the probability density function of a random variable with geometric distribution:

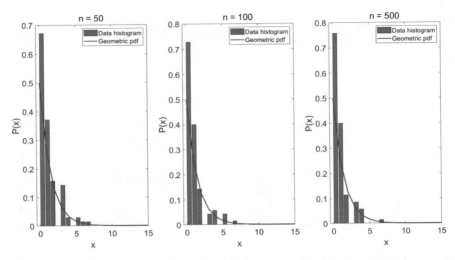

Fig. 2.3 The connection between the Bernoulli distribution and the geometric distribution

```
hold on
% Computation of the geometric probability density function:
pd = geopdf(0:n,p);
% Graphical representation of the probability density function:
plot(1:30,pd(1:30), 'r')
% Labels
xlabel('x')
ylabel('P(x)')
% Legend
legend ('Histogram', 'pdf')
```

Starting from the previous code, Fig. 2.3 can be obtained by choosing $p = 0.5$, $n = \begin{bmatrix} 50 & 100 & 500 \end{bmatrix}$ and $m = 100$. It can be noticed that by changing the number of independent Bernoulli events, the moment when the first value equaling 1 occurs, leads to a distribution that is closer to the geometric distribution.

2.5.5 The Central Limit Theorem

To illustrate the central limit theorem, we consider the case where $X_1, \ldots, X_n$ are independent random variables, uniformly distributed over the range [0, 1]. The purpose of the application is to show the influence of the sample's length n on the distribution of the sum $S = X_1 + \cdots + X_n$. Different values are tested for n, more precisely 1, 2, 10, 30.

```matlab
clear all
close all
clc

% Number of samples:
m = 1000;
% Length of each sample:
n = [1 2 10 30];


figure
for  i = 1:length(n)
    % Generation of m samples having n values with a uniform
distribution U(0,1)
    X = rand(m,n(i));
    % Samples sum
    S = sum(X,2);


    % Computation of the histogram:
    [h, bin] = hist(S);
    dx = diff(bin(1:2));
    h = h/sum(h*dx);
    % Graphical representation of the histogram:
    subplot(1,length(n),i)
    bar(bin,h);
    hold on


    % Normal distribution according to the central limit theorem:
    mu = n(i)/2;
    sigma = sqrt(n(i)*1/12);
    x = 0:0.001:n(i)
    pdf = normpdf(x,mu,sigma);
```

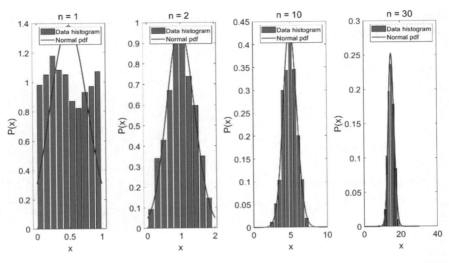

Fig. 2.4 Illustration of the central limit theorem for the sum of n random variables with uniform distribution in the range [0, 1]

```
% Graphical representation of the normal distribution
hold on
plot(x,pdf, 'r')
% Labels
xlabel('x')
ylabel('P(x)')
% Legend
legend ('Histogram', 'pdf')
% Figure title
title(['n = ', num2str(n(i))]);
end
```

Figure 2.4 shows the results of the previous code. If $n = 1$, the uniform distribution is obtained. If n increases, the sum can be better approximated by the normal distribution.

2.5.6 Problem

Let consider a quiz, consisting of 4 questions. Each question has 2 answers, one correct and one incorrect. Determine the probability mass function (PMF) and the probability distribution function (PDF) of the binomial random variable associated to this test. Draw the graphics of these two functions and calculate the likelihood that, with a random choice of answers, the number of correct answers will be at least equal to 2.

```matlab
clear all
close all
clc

% Number of independent events:
n = 4;
% Number of possible successes:
k = 0:n;
% Probability of success:
p = 1/2;

% PMF
p_k = binopdf(k,n,p);

% Graphical representation
figure;
stem(k, p_k, 'o', 'MarkerFaceColor', 'blue', 'LineWidth', 1.5)
xlabel('k')
set(gca,'XTick',k)
ylabel('P(k)')
title('PMF')

% PDF
% Probability distribution function for the binomial random variable X
F_k = binocdf(k,n,p);

% Add two new values on the plot:
k_grafic = [min(k)-1, k, max(k)+1];
F_k_grafic = [0, F_k, 1];

% Graphical representation
figure(3);
stairs(k_grafic, F_k_grafic, 'LineWidth', 1.5)
```

```
hold on

plot(k, F_k, 'o', 'MarkerFaceColor', 'blue', 'MarkerSize', 5)

xlabel('k')

ylabel('F_X(k)')

title('PDF')


% P{2 <= X <= 4}

P_2_4 = binocdf(4,n,p) - binocdf(2,n,p) + p_k(3);
```

2.6   Tasks

Make a report that contains the answers to the following exercises:

1. Using the Probability Distribution Tool, visualize the probability distribution functions and probability density functions for the binomial distribution, Poisson distribution, normal distribution, and uniform distribution. Analyze the influence of the parameters on these functions.
2. Study the connections that exist between the Bernoulli distribution, binomial distribution and geometric distribution, using the code in this paper.
3. Simulate n realizations of the coin-toss experiment, knowing that the successful event (obtaining the head) has the probability equaling p.
4. Solve the problem from Sect. 2.5.6 and correlate the results with those obtained by running the code.
5. Modify the code in Sect. 2.5.5 in order to illustrate the central limit theorem for the mean value of n random variables uniformly distributed over the range [0, 1]. Analyze the influence of n on the results.
6. Modify the code in Sect. 2.5.3 in order to obtain the graph in Fig. 2.2. Comment the result.

2.7   Conclusions

This work is focused on the study of probability distribution functions with examples of applications concerning the communication systems. The central limit theorem is also introduced. For the application part, Matlab functions are provided to facilitate the study of distributions. These functions are then used in examples to highlight the connection that exists between different distributions and to illustrate the central limit theorem.

References

1. Lee JK (2010) Statistical bioninformatics. John Wiley & Sons Inc., For Biomedical and Life Science Researchers
2. Ewens WJ, Grant GR (2005) Statistical methods in bioinformatics. Springer- Verlag, An introduction
3. Stark H, Woods JW (2002) Probability and random processes with applications to signal processing. Prentice-Hall
4. Matlab statistics and machine learning toolbox

Chapter 3
Joint Random Variables

3.1 Definition and Characteristics

Definition: A joint random variable is a random variable composed of two random variables X and Y [1].

Question: Why are joint random variables necessary?

Answer: Joint random variables are needed in order to make predictions about the random variable Y, based on the observations made on the random variable X. Two extreme situations are possible:

- X and Y are *independent*: the observation of X, do not give any information about Y.
- X and Y are *linear dependent*: $Y = aX + b$, so, the observation of X immediately gives Y.

Real cases are between the two situations presented earlier.

Joint random variables are used in experiments described by two random variables at the same time. In this case, the simultaneous characterization of these variables is necessary, while their independent analysis does not offer enough information on the experiment.

The weight and the volume of a sample of water [2], or the communication systems [3] can be modeled by joint random variables.

Example

Like random variables, joint random variables can be described by means of characteristic functions and values.

3.2   Characteristic Functions

3.2.1   Joint Probability Distribution Function

Definition: Let consider the joint random variables X and Y. The *joint probability distribution function* is denoted by $F_{X,Y}(x, y)$ and it is defined as being the function $F_{X,Y} : \mathbb{R}^2 \to [0, 1]$ with the following expression:

$$F_{X,Y}(x, y) = P\{X \leq x, Y \leq y\}, \forall x, y \in \mathbb{R}.$$

Properties:

1. $F_{X,Y}(-\infty, -\infty) = 0$;
2. $F_{X,Y}(\infty, \infty) = 1$;
3. $F_{X,Y}(-\infty, y) = F_{X,Y}(x, -\infty) = 0$;
4. $F_{X,Y}(x, \infty) = F_X(x)$ and $F_{X,Y}(\infty, y) = F_Y(y)$;
5.   If $x_1 \leq x_2$ și $y_1 \leq y_2$, then $F_{X,Y}(x_1, y_1) \leq F_{X,Y}(x_2, y_2)$;
6. $P\{x_1 < X \leq x_2, y_1 < Y \leq y_2\} \quad = \quad F_{X,Y}(x_2, y_2) \quad - \quad F_{X,Y}(x_2, y_1) \quad - \\ -F_{X,Y}(x_1, y_2) + F_{X,Y}(x_1, y_1)$.

3.2.2   Joint Probability Density Function

Definition: Let consider the joint random variables X and Y, characterized by the *joint probability density function* denoted by $f_{X,Y}(x, y)$. If the joint probability distribution function $F_{X,Y}(x, y)$ is continuous and differentiable, then the joint probability density function is defined as:

$$f_{X,Y}(x, y) = \frac{\partial^2 F_{X,Y}(x, y)}{\partial x \partial y}.$$

Properties:

1. $f_{X,Y}(x, y) \geq 0, \forall x, y \in \mathbb{R}$;
2. $\int\limits_{-\infty}^{\infty} \int\limits_{-\infty}^{\infty} f_{X,Y}(x, y)dxdy = 1$.

Two random variables X and Y are independent, if the events $\{X \leq x\}$ and $\{Y \leq y\}$ are independent $\forall x, y \in \mathbb{R}$. In this case [4]:
$$F_{X,Y}(x, y) = F_X(x)F_Y(y) \text{ and } f_{X,Y}(x, y) = f_X(x)f_Y(y).$$

Remark

3.2.3 Marginal Density Functions

The following marginal density functions can be defined for the joint random variables X and Y [5]:

1. $f_X(x) = \int\limits_{-\infty}^{\infty} f_{X,Y}(x, y)dy;$

2. $f_Y(y) = \int\limits_{-\infty}^{\infty} f_{X,Y}(x, y)dx.$

3.3 Characteristic Values

In practice, it is important to have some measures characterizing the prediction of Y based on the observation of X. These measures are given by the joint moments.

3.3.1 Joint Moments of Order ij

Definition: Let consider the joint random variables X and Y. The *joint moment of order* ij is denoted by ξ_{ij} and it is defined as being:

$$\xi_{ij} = E\{X^i Y^j\}.$$

- If X and Y are discrete random variables:

$$\xi_{ij} = E\{X^i Y^j\} = \sum_l \sum_m x_l^i y_m^j P_{X,Y}(x_l, y_m),$$

where $P_{X,Y}(x_l, y_m) = P(X = x_l, Y = y_m)$ is the joint probability mass function of X and Y;

- If X and Y are continuous random variables:

$$\xi_{ij} = E\{X^i Y^j\} = \iint\limits_{-\infty}^{\infty} x^i y^j f_{X,Y}(x, y) dx dy.$$

- The mixed second order moment (obtained for $i = j = 1$) is called the *correlation* of the two random variables X and Y: $R_{X,Y} = E\{XY\}$.
- If $R_{X,Y} = E\{XY\} = 0$, then the random variables X and Y are *orthogonal*.

Remark

3.3.2 Joint Centered Moments of Order ij

Definition: Let consider the joint random variables X and Y. The *joint centered moments of order ij* are denoted by m_{ij} and they are given by:

$$m_{ij} = E\{(X - E\{X\})^i (Y - E\{Y\})^j\},$$

where $E\{X\}$ and $E\{Y\}$ are expectations of variables X and Y.

- The mixed centered moment of second order (obtained for $i = j = 1$) is called the *covariance* of the two random variables X and Y:
$Cov\{X, Y\} = E\{(X - E\{X\})(Y - E\{Y\})\}$.
- By taking into consideration only one random variable X, the *autocovariance* can be defined:

Remark $Cov\{X, X\} = E\{(X - E\{X\})^2\} = \sigma_X^2$,

where σ_X^2 is the dispersion of variable X.

The covariance represents a measure of the strength of the relation existing between the random variables X and Y, taking both positive and negative values. If:

- $Cov\{X, Y\} > 0$, then large values of X correspond to large values of Y, or small values of X correspond to small values of Y.
- $Cov\{X, Y\} < 0$, then large values of X correspond to small values of Y, or otherwise.

Another measure of the strength of the relation between the random variables X and Y is obtained by normalizing the covariance between $[-1, 1]$ and it is called the *correlation coefficient* [6].

3.3.3 Correlation Coefficient Between X and Y

Definition: Let consider the joint random variables X and Y, having the corresponding dispersions σ_X^2 and σ_Y^2. The *correlation coefficient* between X and Y is denoted by $\rho_{X,Y}$ and it is defined as:

$$\rho_{X,Y} = \frac{Cov\{X, Y\}}{\sqrt{\sigma_X^2 \sigma_Y^2}}.$$

Property: $-1 \leq \rho_{X,Y} \leq 1$.

- If $Cov\{X, Y\} = 0$, then $\rho_{X,Y} = 0$, and the random variables X şi Y are *uncorrelated*.

Remark

3.3.4 Correlation Matrix

Definition: The *correlation matrix* of the random variables X and Y is denoted by C and it is defined as:

$$C = \begin{bmatrix} Cov\{X, X\} & Cov\{X, Y\} \\ Cov\{Y, X\} & Cov\{Y, Y\} \end{bmatrix} = \begin{bmatrix} \sigma_X^2 & \rho_{X,Y}\sigma_X\sigma_Y \\ \rho_{X,Y}\sigma_X\sigma_Y & \sigma_Y^2 \end{bmatrix},$$

where σ_X^2 and σ_Y^2 are the dispersions of variables X and Y and $\rho_{X,Y}$ is the correlation coefficient.

In practice, each element of matrix C indicates how the two random variables vary.

3.4   Sample Covariance and Correlation

3.4.1   Definition

When samples are considered, the covariance and the correlation coefficient can be computed by using the previously introduced formulas, adapted for discrete random variables. The new relations are detailed next.

Let consider $x = \{x_1, \ldots, x_N\}$ and $y = \{y_1, \ldots, y_N\}$ two samples having N elements. The covariance between x and y is given by:

$$cov(x, y) = \frac{1}{N-1} \sum_{i=1}^{N} (x_i - \overline{x})(y_i - \overline{y}),$$

while the correlation coefficient is:

$$r(x, y) = \frac{\sum_{i=1}^{N} (x_i - \overline{x})(y_i - \overline{y})}{\sqrt{\sum_{i=1}^{n} (x_i - \overline{x})^2 \sum_{i=1}^{n} (y_i - \overline{y})^2}},$$

where $\overline{x} = \frac{1}{N} \sum_{i=1}^{N} x_i$ and $\overline{y} = \frac{1}{N} \sum_{i=1}^{N} y_i$ are the mean values of vectors x and y.

The correlation matrix C can be computed by using the relation presented in the previous section.

3.4.2   Matlab Functions

The covariance and the correlation coefficient can be easily computed using Matlab. For this purpose, two predefined functions can be applied [7]:

- The covariance can be determined by using *cov* function, which computes the correlation matrix C, and next it extracts the samples covariance.
- The correlation coefficient can be computed by using the *corrcoef* function which gives the following R matrix:

$$R = \begin{bmatrix} r(x, x) & r(x, y) \\ r(y, x) & r(y, y) \end{bmatrix}.$$

- `corrcoef` function returns normalized values.
- `corrcoef` function returns NaN if all the elements in a vector are identical.
- The elements on the main diagonal of R are equaling 1.

Remark

3.5 Applications and Matlab Examples

3.5.1 Joint Random Variables

Exercise 1: The joint PMF of a binary channel is $P(XY) = \begin{bmatrix} 1/4 & 1/4 \\ 1/4 & 1/4 \end{bmatrix}$. Under the assumption of a polar signaling $X = \{-1, +1\}$ and $Y = \{-1, +1\}$. Compute the covariance matrix and the correlation coefficient. Compare the results with those obtained by using the information theory.

Solution:

The correlation coefficient is defined as:

$$\rho_{X,Y} = \frac{Cov\{X, Y\}}{\sqrt{\sigma_X^2 \sigma_Y^2}},$$

so, the covariance $Cov\{X, Y\}$ and the dispersions σ_X^2 and σ_Y^2 are needed.

Thus, the first step in order to solve this exercise is to compute the covariance as:

$$Cov\{X, Y\} = E\{(X - E\{X\})(Y - E\{Y\})\}.$$

Its value depends on the probabilities $P(X) = [p_i]$ and $P(Y) = [q_j]$, that can be obtained starting from the joint PMF, as the sum of the elements on each line and column. Therefore:

$$P(X) = \begin{bmatrix} 1/2 \\ 1/2 \end{bmatrix}$$

and

$$P(Y) = \begin{bmatrix} 1/2 & 1/2 \end{bmatrix}.$$

Knowing that:

$$E\{X\} = \sum_i x_i p_i = -1\frac{1}{2} + 1\frac{1}{2} = 0$$

and

$$E\{Y\} = \sum_j y_j q_j = -1\frac{1}{2} + 1\frac{1}{2} = 0,$$

then the covariance becomes:

$$Cov\{X, Y\} = E\{XY\} = \sum_i \sum_j x_i y_j p_{ij} =$$

$$= (-1)(-1)\frac{1}{4} + (-1)1\frac{1}{4} + (1)(-1)\frac{1}{4} + (1)(1)\frac{1}{4} = 0,$$

where $P(XY) = [p_{ij}]$.

If $Cov\{X, Y\} = 0$, the correlation coefficient is $\rho = 0$, so X and Y are uncorrelated.

The independence of X and Y is obtained if:

$$p_{i/j} = p_i = \frac{1}{2}$$

$$q_{j/i} = q_j = \frac{1}{2}$$

In this case, the channel is independent. The covariance matrix is given by:

$$C = \begin{bmatrix} Cov\{X, X\} & Cov\{X, Y\} \\ Cov\{Y, X\} & Cov\{Y, Y\} \end{bmatrix}.$$

For this matrix:

$$Cov\{X, Y\} = Cov\{Y, X\} = 0,$$

and the autocovariances are:

$$Cov\{X, X\} = \sigma_X^2 = \sum_i (x_i - E\{X\})^2 p_i = (-1 - 0)^2\frac{1}{2} + (1 - 0)^2\frac{1}{2} = 1$$

and

$$Cov\{Y, Y\} = \sigma_Y^2 = \sum_j (y_j - E\{Y\})^2 q_j = (-1 - 0)^2 \frac{1}{2} + (1 - 0)^2 \frac{1}{2} = 1.$$

By using the previously obtained values, the covariance matrix will be:

$$C = \begin{bmatrix} 1 & 0 \\ 0 & 1 \end{bmatrix}.$$

Matlab code:

```
clear all
close all
clc

X = [-1, 1];
Y = [-1, 1];
P_XY = [1/4 1/4; 1/4 1/4];

P_X = sum(P_XY,2);
P_Y = sum(P_XY,1);

mu_X = sum(X.*P_X');
mu_Y = sum(Y.*P_Y);

XY = [X(1)*Y; X(2)*Y];
cov_XY = sum(sum(XY.*P_XY));

disp_x = sum((X-mu_X).^2.*P_X');
disp_y = sum((Y-mu_Y).^2.*P_Y);

rho= cov_XY/(sqrt(disp_x*disp_y));

C = [disp_x cov_XY; cov_XY disp_y];
```

Exercise 2: Let consider a binary symmetric channel. If the signaling is polar with $X = \{-1, +1\}$ and $Y = \{-1, +1\}$ and that the channel is used to capacity, compute the correlation matrix and the correlation coefficient. Discuss the results for $p = 10^{-3}$, $p = 10^{-1}$ and $p = 0.5$.

Solution:

The correlation coefficient is given by:

$$\rho_{X,Y} = \frac{Cov\{X, Y\}}{\sqrt{\sigma_X^2 \sigma_Y^2}},$$

where σ_X^2 and σ_Y^2 are the dispersions of X and Y, while the covariance is computed as:

$$Cov\{X, Y\} = E\{(X - E\{X\})(Y - E\{Y\})\}.$$

To obtain the covariance, the probabilities $P(X) = [p_i]$ are $P(Y) = [q_j]$ are needed. These values can be obtained knowing that a symmetric binary channel is considered and that it is used to capacity.

For this type of channel, the noise matrix has the following expression:

$$P(Y|X) = \begin{bmatrix} 1-p & p \\ p & 1-p \end{bmatrix},$$

while the capacity is given by:

$$C = 1 + p \mathrm{ld} p + (1-p) \mathrm{ld}(1-p).$$

This expression is obtained for:

$$P(X) = [p_i] = \begin{bmatrix} 1/2 \\ 1/2 \end{bmatrix}$$

and

$$P(Y) = [q_j] = \begin{bmatrix} 1/2 & 1/2 \end{bmatrix}.$$

Therefore, the mean values of the two variables are:

$$E\{X\} = \sum_i x_i p_i = -1\frac{1}{2} + 1\frac{1}{2} = 0,$$

$$E\{Y\} = \sum_j y_j q_j = -1\frac{1}{2} + 1\frac{1}{2} = 0,$$

and the corresponding dispersions are:

$$\sigma_X^2 = \sum_i (x_i - E\{X\})^2 p_i = (-1 - 0)^2 \frac{1}{2} + (1 - 0)^2 \frac{1}{2} = 1,$$

$$\sigma_Y^2 = \sum_j (y_j - E\{Y\})^2 q_j = (-1 - 0)^2 \frac{1}{2} + (1 - 0)^2 \frac{1}{2} = 1.$$

By taking into consideration all the previously obtained results, the correlation coefficient becomes:

$$\rho_{X,Y} = Cov\{X, Y\},$$

while the covariance is given by:

$$Cov\{X, Y\} = E\{XY\} = \sum_i \sum_j x_i y_j p_{ij}.$$

In this expression, $[p_{ij}]$ are the elements of the joint probability matrix $P(XY)$:

$$P(XY) = P(X)P(Y|X) = \begin{bmatrix} 1/2 & 0 \\ 0 & 1/2 \end{bmatrix} \begin{bmatrix} 1-p & p \\ p & 1-p \end{bmatrix} = \begin{bmatrix} 1/2(1-p) & 1/2p \\ 1/2p & 1/2(1-p) \end{bmatrix}.$$

In the end, the covariance is computed as:

$$Cov\{X, Y\} = (-1)(-1)\frac{1}{2}(1-p) + (-1)(1)\frac{1}{2}p + (1)(-1)\frac{1}{2}p$$
$$+ (1)(1)\frac{1}{2}(1-p) = 1 - 2p.$$

Discussion:

- If the value of p increases, the covariance will decrease.
- If the value of p decreases, the covariance will increase.
- At limit, the covariance is zero and the variables X and Y are uncorrelated.

Different values of p:

- For $p = 10^{-1}$:

$$Cov\{X, Y\} = 1 - 2 \times 10^{-1} = 0.8;$$

$$\rho_{X,Y} = Cov\{X, Y\} = 0.8;$$

$$C = \begin{bmatrix} \sigma_X^2 & \rho_{X,Y}\sigma_X\sigma_Y \\ \rho_{X,Y}\sigma_X\sigma_Y & \sigma_Y^2 \end{bmatrix} = \begin{bmatrix} 1 & 0.8 \\ 0.8 & 1 \end{bmatrix}$$

- For $p = 10^{-3}$:

$$Cov\{X, Y\} = 1 - 2 \times 10^{-3} = 0.998;$$

$$\rho_{X,Y} = Cov\{X, Y\} = 0.998;$$

$$C = \begin{bmatrix} \sigma_X^2 & \rho_{X,Y}\sigma_X\sigma_Y \\ \rho_{X,Y}\sigma_X\sigma_Y & \sigma_Y^2 \end{bmatrix} = \begin{bmatrix} 1 & 0.998 \\ 0.998 & 1 \end{bmatrix}$$

- For $p = 0.5$:

$$Cov\{X, Y\} = 1 - 2 \times 0.5 = 0;$$

$$\rho_{X,Y} = Cov\{X, Y\} = 0 \Rightarrow X \text{ and } Y \text{ are uncorrelated.}$$

$$C = \begin{bmatrix} \sigma_X^2 & \rho_{X,Y}\sigma_X\sigma_Y \\ \rho_{X,Y}\sigma_X\sigma_Y & \sigma_Y^2 \end{bmatrix} = \begin{bmatrix} 1 & 0 \\ 0 & 1 \end{bmatrix}.$$

3.5.2   Sample Correlation

The correlation between two samples can be used for message decoding in a transmission system. Let r be the received message. Knowing that the messages are encoded using the code denoted by $C(n, m)$, the correct codeword corresponding to r, can be obtained by computing the correlation coefficient between the received binary sequence and all the possible codewords. The correct codeword is the one for which the correlation coefficient has the largest value.

Exercise: Let consider the cyclic one error correcting Hamming code having $k = 3$ control symbols and the received word $r = \begin{bmatrix} 0 & 0 & 1 & 0 & 0 & 0 & 1 \end{bmatrix}$. Based on the above theoretical part, determine the corresponding correct word using Matlab functions.

Matlab code:

```
clear all
close all
clc
% Number of control symbols
k = 3;
% Generate the primitive polynomial of degree k = 3
pol = gfprimdf(k);
% Display the obtained polynomial
gfpretty(gfprimdf(k));
% Generate the control matrix H and the code generating
% matrix G, compute the code length n and the length of
% the information sequence m
[H,G,n,m] = hammgen(k,pol);
% Verify that H and G are orthogonal
mod(G*H',2);
% Generate all messages having m bits (as characters)
x = dec2bin(0:2^m-1);
% Transform the characters into numbers
for i=1:2^m
    for j=1:m
        x_num(i,j)=str2num(x(i,j));
    end
end
% Code all the messages
enc_x = encode(x_num,n,m,'Hamming/binary');
% The received message
r = [0 0 1 0 0 0 1]
% Determine the correlation coefficient between the received % message
and all the possible codewords
for i = 1:size(enc_x,1)
    temp = corrcoef(r,enc_x(i,:));
    cor(i,1) = temp(1,2);
end
% Select the maximum value of the correlation coefficient
[max_cor]=max(cor)
% Determine the codewords corresponding to the previously obtained
positions
cuv_corect = enc_x(find(cor == max_cor),:)
```

3.6 Tasks

Make a document containing the solution to the following exercises:

1. Study the relation between the mathematical solution and the Matlab code given for exercise 1 in Sect. 3.5.1. Add comments for each code line. Explain why the Matlab functions corrcoef and cov cannot be directly applied?
2. Write the Matlab code for exercise 2 in Sect. 3.5.1, by using the code given for the first exercise.
3. For the exercise in Sect. 3.5.2, consider $r = \begin{bmatrix} 1 & 0 & 1 & 0 & 0 & 0 & 1 \end{bmatrix}$ and $r = \begin{bmatrix} 0 & 0 & 0 & 0 & 0 & 0 & 1 \end{bmatrix}$. What does it happen in these cases? Explain the results.

3.7 Conclusions

This work studies the joint random variables and their use in modeling binary transmission systems. Both theoretical aspects (definitions, characteristics, properties) and Matlab implementation details are introduced.

References

1. Naforniță I (2008) Teoria estimării bazată pe model. Editura Politehnica, Timișoara
2. Lee JK (2010) Statistical bioninformatics. John Wiley & Sons Inc., For Biomedical and Life Science Researchers
3. Alexandru Spătaru (1996) Semnale și perturbații. Editura Tehnică
4. Ciuc M, Vertan C (2005) Prelucrarea Statistică a Semnalelor. MatrixRom
5. Stark H, Woods JW (2002) Probability and random processes with applications to signal processing. Prentice-Hall
6. Kay SM (1993) Fundamentals of statistical signal processing- estimation theory, vol I. Prentice-Hall
7. Matlab statistics and machine learning toolbox

Chapter 4
Random Processes

4.1 Definition and Classification

A random process has two properties:

(1)   The samples s_i of the experiment are functions of time (waveforms) and are not real numbers.
(2)   The samples $s_i(t)$ are random in the sense that the waveforms $s_i(t)$ can not be predicted before the experiment.

Let S be the set of events $S = \{S_1, S_2, \ldots, S_n\}$ (Fig. 4.1).

Random process.

A random process $X(t, s)$ is a function defined on S, the samples set and in which each sample s_i is a function of time $s_i(t)$.

- For a random variable, a real number is assigned to each realization
- For a random process, any realization is associated with a waveform (function of time)—the realization/function sample of the random process.

Remark

1. $X(t, s) : S = \{s_i\} \rightarrow \{x_i(t)\}$
- If the time $t = t_k$ is set, the random process becomes a random value

$$X(t, s) : S = \{s_i\} \rightarrow R = \{x_i(t)\}$$

2. A random character is given by characteristic functions:

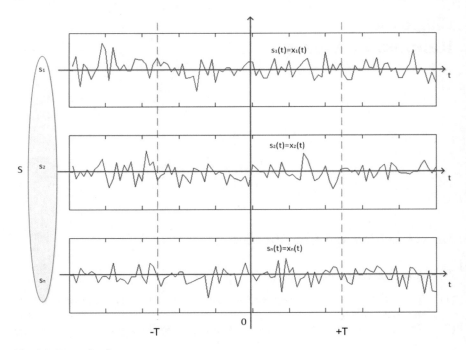

Fig. 4.1 Example of processes

- Distribution function:

$$f_{X^n} = (x_1, x_2, \ldots, x_n; t_1, t_2, \ldots, t_n) = F_{X(t_1),\ldots,X(t_n)}(x_1, x_2, \ldots, x_n)$$

- Probability density function:

$$F_{X^n} = (x_1, x_2, \ldots, x_n; t_1, t_2, \ldots, t_n) = f_{X(t_1),\ldots,X(t_n)}(x_1, x_2, \ldots, x_n)$$

Remark

Most physical phenomena can be modeled as ongoing in time [1], for example, the amount of rainfall in the summer in Cluj-Napoca. The physical process has been ongoing for a long time and will certainly happen from now on. It is therefore convenient to study the probabilistic features of this phenomenon. For this, let $S(n)$ be a random variable that denotes the annual amount of rain during summer. Thus, we obtain an infinite number of random values $(\ldots, S(-1), S(0), S(1), \ldots)$ where the corresponding year for $n = 0$ can be chosen for convenience. A meteorologist could study this phenomenon and determine if the amount of rain increases with the passage of time or if the average amount of rain is constant or the likelihood that the next year the amount of rain will be higher or lower than the current year. In other words, this is a prediction problem, a fundamental problem in many disciplines.

Classification of random processes:

A random process can be:

- Random process continuous in time with:

  - continuous values (e.g. thermal noise $x_i(t)$)
  - discrete values (e.g. random sequences)

- Random process discrete in time with:

  - continuous values (e.g. PAM—pulse amplitude modulation—output signal of a sampling block $x_i(t_i)$)
  - discrete values (e.g. output signal of a quantization block $x_{i_q}(t_i)$).

Random process continuous in time with continuous values—the signal can take as a "continuous" value for "continuous" time. e.g. thermal noise $x_i(t)$ (Fig. 4.2).

Example

Random process continuous in time with discrete values—the signal can take only discrete values and the parameter t has continuous values. e.g. random sequences (Figs. 4.3 and 4.4).

Example

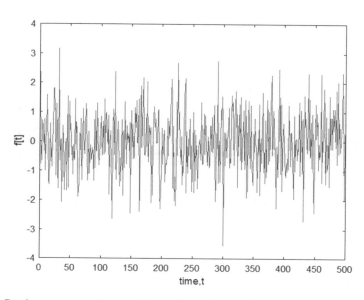

Fig. 4.2 Random process continuous in time with continuous values

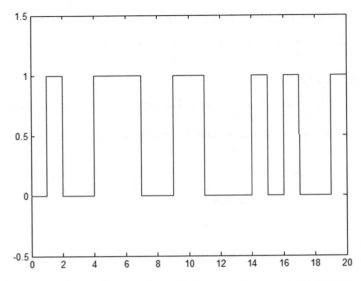

Fig. 4.3 Random process continuous in time with discrete values—random sequences

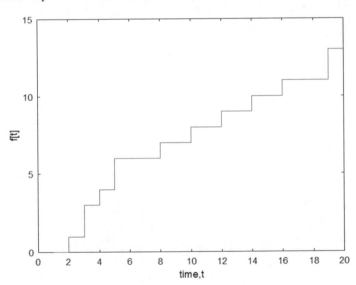

Fig. 4.4 Random process continuous in time with discrete values—random binomial process

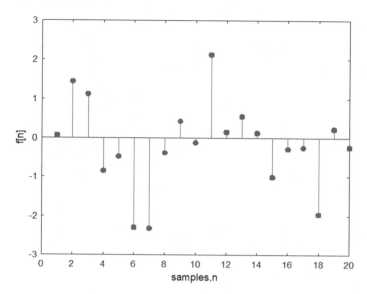

Fig. 4.5 Random process discrete in time with continuous values—Gaussian random process

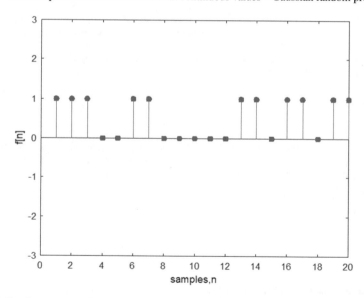

Fig. 4.6 Random process discrete in time with discrete values—Bernoulli random process

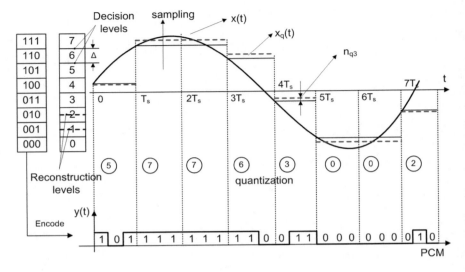

Fig. 4.7 Quantization noise in PCM (from [5])

Random process discrete in time with continuous values—the signal can take a "continuous" value only at discrete moments. e.g. PAM—pulse amplitude modulation—output signal of a sampling block $x_i(t_i)$ (Fig. 4.5).

Example

Random process discrete in time with discrete values—the signal can take discrete values at discrete points in time. e.g. output signal of a quantization block $x_{i_q}(t_i)$) (Figs. 4.6 and 4.7).

Example

4.2   Stationary Process

In the case of random processes, we can observe the time independent start of the observation process, meaning that if the process is divided into a number of time intervals, the different frames (in time) of the process have identical statistical properties [2].

Definition A *stationary process* is a random process that has the same statistical properties at any time ($\tau = t_i - t_j$).

Stationarity is defined in two ways:

(a)   SSS—Strict Sense Stationarity;
(b)   WSS—Weak Sense Stationarity.

Definition A random process $x(t)$ is strict sense stationary if the conditions are met:

1.
$$F_{X^n}(x_1, \ldots, x_n; t_1, \ldots, t_n) = F_{X(t_1), \ldots, X(t_n)}(x_1, \ldots, x_n)$$
$$= F_{X(t_1+\tau), \ldots, X(t_n+\tau)}(x_1, \ldots, x_n), \forall \tau, \forall t_i$$

for any displacement and any possible choice of time t_i. The physical significance is that the distribution function of order n is invariant at the origin of time.

Particular case: $n = 1$

2. $F_{X(t)}(x) = F_{X(t+\tau)}(x), \forall t_n, \tau$

The first order distribution function of a strict sense stationarity random process is independent over time.

Particular case $n = 2$

3. $F_{X(t_1), X(t_2)}(x_1, x_2) = F_{X(t_1+\tau), X(t_2+\tau)}(x_1, x_2), \forall t_n, \tau$

The second order distribution function of a stationary random process depends only on the difference between the observation time and not the particular time in which the random process is observed.

4.2.1 Statistical Mean for a Random Process

(a)   Strict sense stationary random process

Let $X(t)$ a strict sense stationary random process.

Statistical mean of a random process $X(t)$ = statistical mean of a random variable $x(t)$, obtained by considering time t fixed:

$$\mu_X(t) = \{E(X(t))\} = \int_{-\infty}^{+\infty} x f_{X(t)}(x) dx,$$

where $f_{X(t)}(x)$ is the probability density of a random process $x(t)$.

If the random process $X(t)$ is strict sense stationary then $f_{X(t)}(x)$ is independent of time
$$\mu_X(t) = \mu_X = constant, \forall t$$

Remark

The autocorrelation function of a strict sense stationary random process $X(t)$:

$$R_X(t_1, t_2) = E\{X(t_1)X(t_2)\} = \int\limits_{-\infty}^{+\infty}\int\limits_{-\infty}^{+\infty} x_1 x_2 f_{X(t_1)X(t_2)}(x_1 x_2) dx_1 dx_2.$$

If the random process $X(t)$ is strict sense stationary, then $f_{X(t_1)X(t_2)}(x_1 x_2)$ depends only on the difference $t_2 - t_1$
$$R_X(t_1, t_2) = R_X(t_2 - t_1), \forall t_1, t_2$$

Remark

The autocovariance function of a random process $X(t)$:

$$\begin{aligned} C_x(t_1, t_2) &= E\{(X(t_1) - \mu_X)(X(t_2) - \mu_X)\} \\ &= E\{X(t_1)X(t_2) + \mu_X{}^2 - \mu_X[X(t_1) + X(t_2)]\} \\ &= E\{X(t_1)X(t_2)\} + \mu_X{}^2 - \mu_X E\{X(t_1)\} - \mu_X E\{X(t_2)\} \end{aligned}$$

$$C_x = R_X(t_1 - t_2) - \mu_X{}^2 = R_x(\tau) - \mu_X{}^2.$$

(b)   Weak sense stationary random process

Let $X(t)$ a weak sense stationary random process.
   The autocorrelation function of a random stationary process $X(t)$ in a weak sense:

$$R_X(\tau) = E\{X(t + \tau)X(t)\}\forall t, \tau.$$

4.3   Ergodic Processes

The mediations of a random process $X(t)$ can be done in two ways:

(1)   Transversal: fix time t_k and mediate on a sampling collection;
(2)   Along the time axis considering a single realisation $x(t)$.

   Mediating in time on a single realisation we obtain a number and not a time function. If we obtain the same mean on a single realisation (mean in time) with those obtained on a collection of realisations (the statistical mean) we call it a stationary random process.
   Let $X(t)$ be a weak stationary random process and $x(t)$ a single realisation.
   Temporal mean (the continuous component) or the sample mean of $x(t)$ on the interval $[-T, T]$

$$\mu_X(T) = \frac{1}{2T}\int\limits_{-T}^{+T} x(t)dt = \overline{x(t)}$$

$\mu_X(T)$ is a random variable because it depends on the observation interval $[-T, T]$ and the sample $x(t)$. Repeating the mediation experiment on another realisation $\mu_X(T)$ will be different.

Remark

The statistical mean of the mean sample can be calculated:

$$E\{\mu_X(T)\} = E\left\{\frac{1}{2T}\int_{-T}^{+T} x(t)dt\right\} = \frac{1}{2T}\int_{-T}^{+T} E\{x(t)\}dt = \frac{1}{2T}\int_{-T}^{+T} \mu_X dt = \mu_X$$

$$E\{\mu_X(T)\} = \mu_X$$

The temporal mean is an estimation without displacement (without systematic error) of the statistical average μ_X.

Autocorrelation function over time:

$$R_x(\tau, T) = \frac{1}{2T}\int_{-T}^{+T} x(t + \tau)x(t)dt$$

The autocorrelation function over time is a random variable that depends on the observation time T and the realization $x(t)$.

4.4 Autocorrelation Function of a Sum of 2 Independent Processes

Let $s(t), r(t)$ two independent random processes, the sum of the two:

$$r(t) = s(t) + n(t)$$

Autocorrelation function of the sum of the two independent processes:

$$R_r(\tau, T) = \frac{1}{2T}\int_{-T}^{+T} r(t + \tau)r(t)dt = \frac{1}{2T}\int_{-T}^{+T} [s(t + \tau) + n(t + \tau)][s(t) + n(t)]dt$$

$$= \frac{1}{2T}\int_{-T}^{+T} s(t + \tau)s(t)dt + \frac{1}{2T}\int_{-T}^{+T} n(t + \tau)n(t)dt + \frac{1}{2T}\int_{-T}^{+T} s(t + \tau)n(t)dt$$

$$+ \frac{1}{2T} \int_{-T}^{+T} n(t + \tau)s(t)dt$$

$$= R_s(\tau, T) + R_n(\tau, T) + R_{ns}(\tau, T) + R_{sn}(\tau, T)$$

Considering that $R_{ns}(\tau, T)$ and $R_{sn}(\tau, T)$ are 0 based on the signal and noise independence results:

$$R_r(\tau) = R_s(\tau) + R_n(\tau)$$

In the origin $R_r(0) = R_s(0) + R_n(0)$ with physical significance: $P_r = P_s + P_n$, where P_r is the power at the reception, P_s is the signal strength, P_n is the noise power.

- If two signals are statistically independent in the signal space, the scalar product $\langle s, n \rangle = 0$ of the two is null
- Reciprocal is not true, if the scalar product of the two signals is zero does **not** mean that they are independent $\langle s, n \rangle = 0 \nRightarrow s, n$ independent, but s, n uncorrelated.

Remark

4.5   Applications and Matlab Examples

4.5.1   Generation of a Random Process

Example no. 1: An example of a time-continuous random process with discrete values is the so-called random process of a telegraph. Let T_1, T_2, T_3 be a sequence of independent random variables, but similarly distributed, each with an exponential distribution:

$$f_T = \lambda e^{-\lambda s} u(s)$$

At any moment, the random signal of the telegraph, $X(t)$ takes one of the two possible values $X(t) = 0$ or $X(t) = 1$. Suppose the process starts at $t = 0$. The signal remains in the same value for a time interval equal to T_1, from this point it passes to the state $X(t) = 1$. This process stays in this state for another time period equal to T_2 and then changes, goes back to the other state. The process continues to change after some time intervals specific to the exponential random variable sequence [3]. The MATLAB code to generate such a process is like this:

```
N=10; % Number of state changes
Fs=100; % Sampling frequency (samples per second)
lambda=1/2; % Changes frequency (changes per second)
X=[];
S=rand(1,N); % Uniform random variables
T=-log(S)/lambda; % Exponential transformation
V=cumsum(T); % State change time
state=0; Nsold=1;
for k=1:N
Nsnew=ceil(V(k)*Fs); % New time of state change
Ns=Nsnew-Nsold; % Number of current samples
% Change interval
X=[X state*ones(1,Ns)];
state=1-state; % State change
Nsold=Nsnew;
end
t=[1:length(X)]/Fs; % Time axis
plot(t,X) % Result display
xlabel('time, t'); ylabel('X(t)')
axis([0 max(t) -0.1 1.1]) % Manual scaling of axis
```

And the result from running the code should look like the graph from Fig. 4.8.

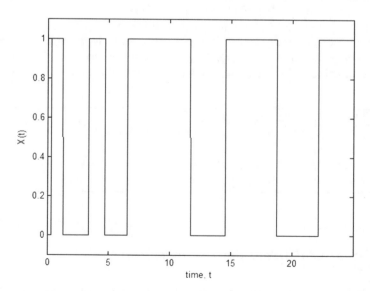

Fig. 4.8 Random process continuous in time with discrete values

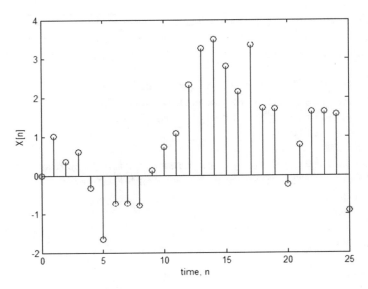

Fig. 4.9 Time discrete random process

Example no. 2: For another time discreet random process we take the following example (Fig. 4.9). Suppose that each experiment produces an independent sequence with a similar distribution of Gaussian random variables of zero mean $W_1, W_2, W_3 \dots$ A discrete time process $X[n]$ could be built as follows:

$X[n] = X[n - 1] + W_n$, with the initial condition $X[0] = 0$

In the process, the value at each point in time is equal to the process value from the previous point plus a random quantity (ascending or descending) that follows a Gaussian distribution [3]. The Matlab code to generate this process is as follows:

```
N=25; % Number of process instances
W=randn(1,N); % Gaussian random variables
X=[0 cumsum(W)]; % Samples of X[n].
stem([0:N],X,'o') % Display the realisation of X[n].
xlabel('time, n'); ylabel('X[n]');
```

4.5.2 Statistical Means for a Random Process

Example no. 3: Let us consider a sinus function with a random frequency $X(t) = \cos(2\pi F t)$, where F is a random variable uniformly distributed over an interval $(0, f_0)$. The statistical mean can be determined as follows:

$$\mu_X(t) = \{E(\cos(2\pi Ft))\} = \frac{1}{f_0} \int\limits_{0}^{f_0} \cos(2\pi ft)df = \frac{\sin(2\pi f_0 t)}{2\pi f_0 t}$$

We can estimate the statistical mean through a Matlab simulation as follows [3]:

```
fo=2; % Maximum frequency
N=1000; % Realisation number
t=[-4.995:0.01:4.995]; % Time axis
F=fo*rand(N,1); % Uniform frequency
x=cos(2*pi*F*t); % Each row is a realisation of pro-
cess sample_mean=sum(x)/N; % Computation of value mean
true_mean=sin(2*pi*fo*t)./(2*pi*fo*t);
plot(t,sample_mean,'-',t,true_mean,'--')    %    Result
display
xlabel('t (secondes)'); ylabel('miu(t)');
```

The result of the simulation looks like in the Fig. 4.10.

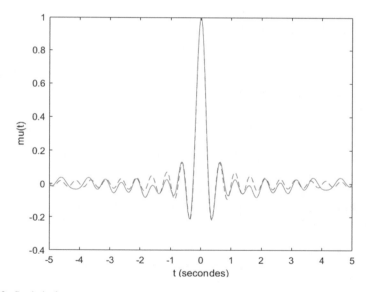

Fig. 4.10 Statistical mean

4.5.3   Ergodic Processes

Example no. 4: For a known ergodic process, the autocorrelation function can be estimated by considering a fairly long-time average for the autocorrelation function of a single realisation.

Let be an ergodic process $X(t) = \sum_{k=1}^{n} \cos(2\pi f_k t + \theta_k)$, where θ_k is independent and with a similar distribution on $(0, 2\pi)$. For some signal $x(t)$ we want to observe the autocorrelation function. If we are given a single realization $x(t)$, which lasts as long as the time interval $(-t_0, t_0)$, then:

$$R_x(t, \tau) = \frac{1}{2t_0 - \tau} \int_{-t_0}^{t_0 - \tau} x(t) x(t + \tau) dt, \text{ for } \tau > 0$$

Using the MATLAB code below we can see the autocorrelation function for this ergodic process [3].

```
N=4 ; % Number of sinusoids
T0=5 ; % Time interval
Ts=0.01 ; % Sampling interval
T=[-t0 :Ts :t0] ; % Time axis
Tau=[-2*t0 :Ts :2*t0] ; % Tau axis
Theta=rand(1,N) ; % Random faze
F=1./[1:N]; % Samples (not random).
X=zeros(size(t));
True_Rxx=zeros(size(tau));
for k=1:N
x=x+cos(2*pi*f(k)*t+2*pi*theta(k)); % Building the pro-
cess
True_Rxx=True_Rxx+cos(2*pi*f(k)*tau)/2 ;   %   Computing
Rxx(tau).
End
subplot(2,1,1); plot(t,x,'-')% Result display
Ylabel('function f');
subplot(2,1,2);plot(tau,True_Rxx,'-')
ylabel('Rx');
```

4.5.4   Autocorrelation Function of a Sum of 2 Independent Processes

Example no. 5: To see how the autocorrelation function of two random processes and their convolution are calculated [4] we have the following example:

```
clc; close all; clear all;
% ========== variables:
fs = 200;
t = linspace(-1,1,fs);
crossCorrelation = zeros(1,length(t));
convolution      = zeros(1,length(t));
set(gcf,'Color',[1 1 1]);
ax = [-1 1 -0.2 1.1];
% ========== plot the signals (y1 and y2):
disp('y1: single rectangle pulse with width of 1.0')
disp('y2: single triangle  pulse with width of 0.5')
y1 = rectpuls(t,1) ;
y2 = tripuls(t,0.5,-1) ;
sub-
plot(4,2,1);plot(t,y1,'Color','blue','LineWidth',2),axi
s(ax);
grid on;
xlabel('t')
ylabel('y1')
sub-
plot(4,2,2);plot(t,y2,'Color','red','LineWidth',2),axis
(ax);
grid on;
xlabel('t')
ylabel('y2')
disp('Press Enter to continue ....');
pause;
% ========== cross correlation of two signals(y1 si
y2):
66print('\n\nCross Correlation of two signals (y1 and
y2):');
iter = length(t);
for i = 1:iter
    moveStep = (2*i-fs)/fs;
    y2_shifted = tripuls(t-moveStep,0.5,-1);
    crossCorrelation(i) = trapz(t, y1.*y2_shifted) ;
    subplot(4,2,3:4)
    hold off;
    plot(t,y1,'Color','blue','LineWidth',2),axis(ax);
    hold on;
```

```
      plot(t,y2_shifted,'Color','red',                       'Lin-
eWidth',2),axis(ax);
    grid on;
    xlabel('t')
    subplot(4,2,5:6)
    hold off

plot(t(1:i),crossCorrelation(1:i),'Color','black','Line
Width',2); axis([-1 1 -0.2 0.5]);
    grid on ;
    xlabel('t')
    ylabel('CrossCorrelation(y1, y2)(t) ') ;
    pause(0.01);
end
disp('Press Enter to continue ....');
pause;
 % ========== convolution of y1 and y2:
  print('\n\nConvolution of two signals (y1 and y2):');
for i = 1:iter
    moveStep = (2*i-fs)/fs;
    y2_shifted = tripuls(-(t-moveStep),0.5,-1);
    convolution(i) = trapz(t, y1.*y2_shifted) ;
    subplot(4,2,3:4)
    hold off;
    plot(t,y1,'Color','blue','LineWidth',2),axis(ax);
    hold on;
    plot(t,y2_shifted,'Color','red'
,'LineWidth',2),axis(ax);
    grid on;
    xlabel('t');
    subplot(4,2,7:8)
    hold off

plot(t(1:i),convolution(1:i),'Color','black','LineWidth
',2); axis([-1 1 -0.2 0.5]);
    grid on ;
    xlabel('t')
    ylabel('Convolution(y1, y2)(t) ')
    pause(0.01)
end
```

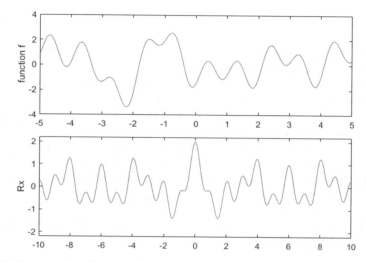

Fig. 4.11 Representation of an ergodic process $X(t)$ and the autocorrelation function

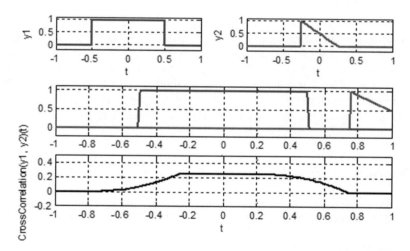

Fig. 4.12 Autocorrelation function of two independent random processes

After running the code above, we will get the Figs. 4.11 and 4.12.

4.6   Tasks

Using Matlab solve the following requirements:

1. Using the code examples above, display and visualize the results and comment on the results.
2. For Example no.5 (sum of two independent processes), visualize the simulation in the paper and comment on the result.
3. Implement the autocorrelation and autocovariance function for a stationary random process using Matlab functions. Show the results.
4. Modify one of the received examples to illustrate a discrete time random process with discrete values. Display the result.
5. Using the received Matlab code, implement a random sinusoidal process with uniform random phases $X(t) = A cos(2\pi f_C t + \theta)$, where A, f_C are constants, and θ is a random variable uniform distributed in the interval $[-\pi, \pi]$, having $f_\theta = \begin{cases} \frac{1}{2\pi}, & -\pi \leq \theta \leq \pi \\ 0, & \hat{i}n\, rest \end{cases}$. Calculate and display the autocorrelation function. What kind of process did you get? A strict sense stationary process or a weak sense stationary process?

4.7 Conclusions

The chapter studies the random processes from the theoretical point of view, giving examples and finally applications. Random processes are defined and classified, and notions about stationary processes and ergodic processes are introduced. You can find examples of implementation of a time continuous process with discrete values, a time discreet random process, an example of computing the statistical mean, an ergodic process, and the calculation of the autocorrelation function.

References

1. Kay SM (2006) Intuitive probability and random processes using MATLAB. Springer
2. Haykin S, Moher M (2009) Communication systems. 5th edn. Wiley
3. Miller S, Childers D (2012) Probability and random processes, 2nd edn. Academic Press, With Applications to Signal Processing and Communications
4. Reza Arfa Github https://github.com/rezaarfa. Accessed 9 September 2021
5. Borda M (2011) Fundamentals in information theory and coding. Springer

Chapter 5
Pseudo-Noise Sequences

5.1 Theoretical Introduction

The pseudo-noise sequences (PNS) are deterministic binary sequences which are like the noise sequences. The noise sequences are characterized by the fact that the probabilities of the appearances of a symbol 0 or 1 are equal and independent of the previous symbols: $p(0) = p(1) = \frac{1}{2}$ [1].

There are a few types of PNS, the most known PNS are the ones of maximum length: $n = 2^k - 1$, generated by feedback shift registers with the characteristic polynomial $g(x)$, which is of degree k and primitive.

The PNSs of maximum length have the following properties:

1. They are periodic sequences of length n.
2. For a period, the number of 1 symbols is with one symbol greater than 0 symbols.
3. A cyclic permutation of a PNS generates a new pseudo-noise sequence, which compared with the original PNS has a number of coincidences smaller with one unit compared with the number of non-coincidences.
4. The modulo-two sum of two pseudo-noise sequences is still a PNS.

5.2 PNS Generator

A PNS of maximum length n can be generated using a Linear Feed-back Shift Register (LFSR) with characteristic polynomial $g(x) = x^k + g_{k-1}x^{k-1} + g_{k-2}x^{k-2} + \cdots + g_1 x + 1$ and a XOR logical gate. The characteristic polynomial is a primitive polynomial with binary coefficients [1]. In Table 5.1 the possible coefficients for primitive polynomials of maximum degree 5 are shown. In the **Appendix A** of this book, the possible coefficients for primitive polynomials of maximum degree 100 are shown.

M. Borda et al., *Randomness and Elements of Decision Theory Applied to Signals*,
https://doi.org/10.1007/978-3-030-90314-5_5

Table 5.1 Coefficients for primitive polynomials of maximum degree 5

k	$a_k a_{k-1} \dots a_0$
1	1 1
2	1 1 1
3	1 0 1 1
4	1 1 0 1
5	1 0 0 1 1
	1 1 0 0 1
	1 0 0 1 0 1
	1 0 1 0 0 1
	1 0 1 1 1 1
	1 1 0 1 1 1
	1 1 1 1 0 1

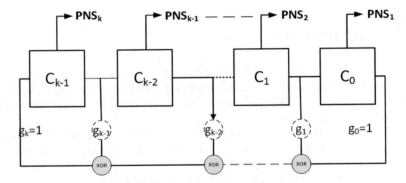

Fig. 5.1 Block scheme of a PNS generator with LFSR and XOR gates

Figure 5.1 shows the circuit for PNS generator which uses:

- a LFSR with k cells
- a XOR gate

The initial state of the register is any state different from the zero state, and the PNS can be obtained from any output of the LFSR.

Problem 5.1 Generate a pseudo-noise sequence of length $n = 7$, using a LFSR with generator polynomial $g(x) = x^3 + x^2 + 1$. The initial state $C_2 C_1 C_0$ is 100 and the PNS is obtained from the cell C_0.

Solution

1. The block scheme of the generator using the polynomial $g(x) = x^3 + x^2 + 1$ Is the one from Fig. 5.2.

Fig. 5.2 Block scheme of
the PNS generator using the
polynomial
$g(x) = x^3 + x^2 + 1$

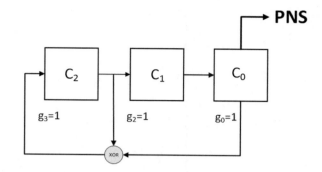

Table 5.2 The evolution of the SR for the above example

T_{CK}	C_2	C_1	C_0	PNS
1	1	0	0	0
2	1	1	0	0
3	1	1	1	1
4	0	1	1	1
5	1	0	1	1
6	0	1	0	0
7	0	0	1	1

2. Starting with the initial state $C_2C_1C_0 = 100$, the register will have all the non-zero states, as it can be seen in Table 5.2.
3. At the outputs of the cells the pseudo-noise sequences are obtained, whit a time delay between them.
4. The final sequence is *0 0 1 1 1 0 1*, for which the probabilities of the symbols *0* and *1* are: $p(0) = \frac{3}{7}$ and $p(1) = \frac{4}{7}$.

5.3 Identification of the Pseudo-Noise Sequences

A pseudo-noise sequence can be identified due to some mathematical functions: autocorrelation function and mutual correlation function [2].

5.3.1 Autocorrelation Function

For a pseudo-noise sequence of length n for which the period of a symbol is T, the autocorrelation function is defined as:

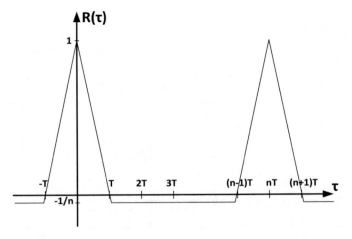

Fig. 5.3 Graph of the autocorrelation function

$$R(\tau) = \frac{1}{nT} \int_0^{nT} PNS(t) \cdot PNS(t - \tau)dt$$

where $PNS(t - \tau)$ is the $PNS(t)$ sequence delayed with τ periods of a symbol.

Because the autocorrelation function is an even function, the relation for the case when $|\tau| \le (n - 1)T$ is:

$$R(\tau) = \begin{cases} \frac{T - |\tau|}{T}, & for |\tau| \le T \\ -\frac{1}{n}, & for T < |\tau| < (n - 1)T \end{cases}$$

The graph of the autocorrelation function can be observed in Fig. 5.3.

By analyzing this function, one can observe that it has a maximum value of 1, when τ is in origin, i.e. $\tau = 0, nT, \ldots$, and a minimum for the remaining of the period.

If we denote by $I_c(\tau)$ the time interval within a period nT during which $PNS(t) = PNS(t - \tau)$, refereed from now on as coincidence interval, and respectively with $I_a(\tau)$ the time interval within a period nT during which $PNS(t) = -PNS(t - \tau)$, refereed from now on as anticoincidence interval, and taking into account the relation $I_c(\tau) + I_a(\tau) = nT$, the autocorrelation function can be written as:

$$R(\tau) = \frac{I_c(\tau) - I_a(\tau)}{nT}$$

For the case when τ represents an integer number of periods of o symbol, $\tau = kT$, the autocorrelation function can be discretized as:

$$R(\tau) = \frac{N_c - N_a}{n}$$

where Nc represents the number of positions for which the sequences $PNS(t)$ and $PNS(t - \tau)$ are identical and Na is the number of the position for which the two sequences are different.

5.3.2 Mutual Correlation Function

Because of the well-marked maximum of the autocorrelation function, there is the possibility that a PNS to be identified during the presence of some numerical errors. The mutual correlation function $R_{12}(\tau)$, computed between correct PNS and PNS with errors, denoted as PNS, is:

$$R_{12}(\tau) = \frac{1}{nT} \int_0^{nT} PNS(t) \cdot PNS'(t - \tau)dt = \frac{I_c(\tau) - I_a(\tau)}{nT}$$

where $I_c(\tau)$ and $I_a(\tau)$ are defined as intervals of coincidence and anticoincidence between PNS and PNS'. For an integer number of periods of a symbol, $\tau = kT$, the mutual correlation function can be discretized as:

$$R_{12}(\tau) = \frac{N_c - N_a}{n}$$

where N_c represents the number of positions for which $PNS(t)$ and $PNS'(t - \tau)$ are identical and N_a is the number of positions for which the above-mentioned two sequences are different.

The property of the autocorrelation function, it has a maximum value when τ is in origin, i.e. $\tau = 0, nT, \ldots$, is maintained also for the mutual correlation function, but only for a maximum number of errors. So, the maximum number of errors for which a PNS can have a maximum value when τ is in origin is:

$$i_{max} < \frac{n + 1}{4}$$

5.4   Applications and MATLAB Examples

5.4.1   PNS Generator Using MATLAB Functions

MATLAB allows the creation of an object *PNSequence* which generates a sequence of pseudorandom binary numbers using a linear-feedback shift register (LFSR). This block implements LFSR using a simple shift register generator.

To generate a PN sequence:

1.   Define and set up your PN sequence object. See Construction.
2.   Call step to generate a PN sequence according to the properties of comm.PNSequence.

The Matlab command H = comm.PNSequence(Name,Value) creates a PN sequence generator object, H, with each specified property set to the specified value. You can specify additional name-value pair arguments in any order as (Name1,Value1,…,NameN,ValueN). The main additional name-value pair arguments are listed in Table 5.3.

Problem 5.2 Using MATLAB instructions, generate a PNS of length n = 7, with an LFSR having the generator polynomial $(x) = x^3 + x^2 + 1$. The initial state is $C_2C_1C_0$: 100.

Solution

```
pns = comm.PNSequence('Polynomial',[3 2 0],  ...
  'SamplesPerFrame',7,'InitialConditions',[1 0 0]);
x = step(pns)
x =

    0
    0
    1
    1
    1
    0
    1
```

Table 5.3 Main additional name-value pair arguments [3]

Polynomial	Generator polynomial Specify the polynomial that determines the shift register's feedback connections. The default is 'z^6 + z + 1'. You can specify the generator polynomial as a string or as a numeric, binary vector that lists the coefficients of the polynomial in descending order of powers. The first and last elements must equal 1, and the length of this vector must be n + 1. The value n indicates the degree of the generator polynomial. Lastly, you can specify the generator polynomial as a numeric vector containing the exponents of z for the nonzero terms of the polynomial in descending order of powers. The last entry must be 0. For example, [1 0 0 0 0 1 0 1] and [8 2 0] represent the same polynomial, g(z) = z8 + z2 + 1 The PN sequence has a period of N = 2n − 1 (applies only to maximal length sequences)
InitialConditions	Initial conditions of shift register Specify the initial values of the shift register as a binary, numeric scalar or a binary, numeric vector. The default is [0 0 0 0 0 1]. Set the vector length equal to the degree of the generator polynomial. If you set this property to a vector, each element of the vector corresponds to the initial value of the corresponding cell in the shift register. If you set this property to a scalar, the initial conditions of all the cells of the shift register are the specified scalar value. The scalar, or at least one element of the specified vector, must be nonzero for the object to generate a nonzero sequence
SamplesPerFrame	Number of outputs per frame Specify the number of PN sequence samples that the step method outputs as a numeric, positive, integer scalar value. The default is 1. If you set this property to a value of M, then the step method outputs M samples of a PN sequence that has a period of N = 2n − 1. The value n represents the degree of the generator polynomial that you specify in the Polynomial property. If you set the BitPackedOutput property to false, the samples are bits from the PN sequence. If you set the BitPackedOutput property to true, then the output corresponds to SamplesPerFrame groups of bit-packed samples

5.4.2   Graphs of Autocorrelation and Mutual Correlation Functions

For a pseudo-noise sequence obtained in MATLAB, one can determine and draw the graphs of the autocorrelation and mutual correlation functions using MATLAB instructions.

(a)   Autocorrelation function

```
vect = [];
for i=0:n-1
  s = circshift(m,i);
  vect = [vect m'*s];
end
vect = vect/n;
function vect=auto_corr(pns)
n=length(pns);
m=(pns-0.5)*2;
```

where pns is the MATLAB variable that contains the PNS, and *vect* is the vector that contains the values of the autocorrelation function.

(b)   Mutual correlation function

```
function vect = mutual_corr(pns,err)
n=length(pns);
e = randerr(1,n,err);
y = double(xor(pns',e));
s1=(pns-0.5)*2;
s2=(y-0.5)*2;
vect = [];
for i=0:n-1
  s = circshift(s2',i);
  vect = [vect s1'*s];
end
vect = vect/n;
```

where *pns* is the MATLAB variable that contains the PNS, *err* is the number of errors that appear in the sequence, and *vect* is the vector that contains the values of the autocorrelation function.

Problem 5.3 Using the above-defined functions, determine and draw the graphs of the autocorrelation and mutual correlation functions for the PNS generated in Problem 5.1, in the cases when there are no numerical errors, there is 1 error and respectively 2 errors.

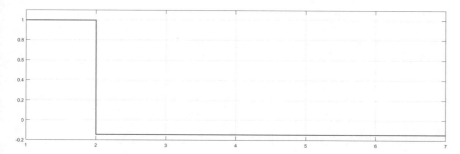

Fig. 5.4 The graph of the autocorrelation function for a PNS of length 7

Solution

(a)   The autocorrelation function is determined for the case when there are no errors.

```
vect=auto_corr(x);
stairs(vect)
```

The graph is the one from Fig. 5.4.

(b)   The mutual correlation function is determined for the case when there are 1 or more numerical errors. For one error the MATLAB instructions are:

```
vect = mutual_corr(x,1);
stairs(vect)
```

The graph of the mutual correlation function is the one from Fig. 5.5.

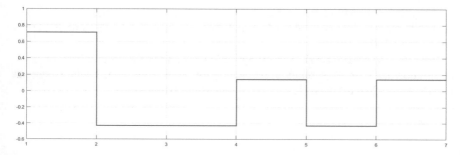

Fig. 5.5 The graph of the mutual correlation function for a PNS of length 7, when 1 error is present

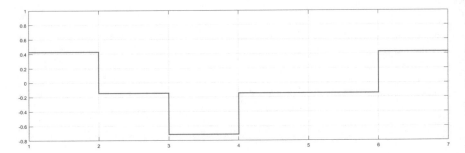

Fig. 5.6 The graph of the mutual correlation function for a PNS of length 7, when 2 errors are present

In Fig. 5.5 one can observe that the maximum is maintained in the origin when 1 error appears. For 2 errors the MATLAB instructions are:

```
vect = mutual_corr(x,2);
stairs(vect)
```

The graph of the mutual correlation function from Fig. 5.6 indicates that we can no longer identify where is the origin of the PNS, due to the appearance of more than 2 errors in the 7th length sequence.

5.5 Tasks

Task 1. Determine and draw the graph of the autocorrelation and the mutual correlation function for the pseudo-noise sequence obtained at Problem 5.1, from paragraph 5.2 in the case when there are no errors, there is one numerical error and there are two numerical errors.

Task 2. Using MATLAB instructions, generate a PNS of length n = 15, with an LFSR having the generator polynomial $(x) = x^4 + x + 1$. The initial state is $C_3C_2C_1C_0$: 1001.

Task 3. Using MATLAB instructions determine and draw the graph of the auto-correlation and mutual correlation function for the PNS from **Task 2**, when there are no errors, or when there are 1, 2, 3 and 4 numerical errors.

5.6   Conclusions

The chapter "Binary pseudo-noise sequence generator" presents the properties of the pseudo noise sequences and the correlation functions. It shows how a PNS generator can be designed in the MATLAB environment.

References

1. Borda M (2011) Fundamentals in information theory and coding. Springer
2. Radu M, Stoica S (1988) Telefonie Numerică, Ed. Militară
3. MATLAB Help www.mathworks.com

Chapter 6
Markov Systems

6.1 Solved Problems

Example 6.1 [1] One classical example of using Markov chains is weather fore-casting. In meteorological stations the weather is observed and classified daily at the same hour, as follows:

$s_1 = S$ (sun).
$s_2 = C$ (clouds).
$s_3 = R$ (rain).

Based on daily measurements, meteorologists computed the transition probability matrix:

$$M = \begin{bmatrix} 0.6 & 0.2 & 0.2 \\ 0.3 & 0.4 & 0.3 \\ 0.2 & 0.2 & 0.6 \end{bmatrix}$$

Some questions raised:

- knowing that today is a sunny day, what is the probability that the forecast for the next seven days is: $s_1\ s_1\ s_2\ s_1\ s_2\ s_3\ s_1$?

  The answer is:

  $$p(s_1) \cdot p(s_1/s_1) \cdot p(s_1/s_1) \cdot p(s_2/s_1)p(s_1/s_2) \cdot p(s_2/s_1) \cdot p(s_3/s_2) \cdot p(s_1/s_3)$$
  $$= 1 \cdot 0{,}6 \cdot 0{,}6 \cdot 0{,}2 \cdot 0{,}3 \cdot 0{,}2 \cdot 0{,}3 \cdot 0{,}2 = 2{,}592 \cdot 10^4$$

- what is the probability that the weather becomes and remains sunny for three days, knowing that the previous day it had rained?

  The answer is:

$$p(s_3) \cdot p(s_1/s_3) \cdot p(s_1/s_1) \cdot p(s_1/s_1) \cdot [1 - p(s_1/s_1)] = 1 \cdot 0{,}2 \cdot 0{,}6 \cdot 0{,}6 \cdot (1 - 0{,}6)$$
$$= 2{,}88 \cdot 10^{-2}$$

Example 6.2 [2] Let the transition matrix for a Markov chain:

$$M = \begin{bmatrix} 0.2 \ 0.3 \ 0.5 \\ 0.4 \ 0.4 \ 0.2 \\ 0.4 \ 0.6 \ \ 0 \end{bmatrix}$$

(a) Draw the graph corresponding to the Markov system;
(b) Determine the probability of transition from state s_2 to state s_3 in 2 steps;
(c) Determine the probability of transition from state s_1 to state s_2 in 3 steps;
(d) Knowing that the initial state of the system is $P^{(0)} = [0.4, \quad 0.1, \quad 0.5]$ determine the state of the system after 3 steps;
(e) Determine the stationary state of the system.

Solution

(a) The diagram of the Markov system:

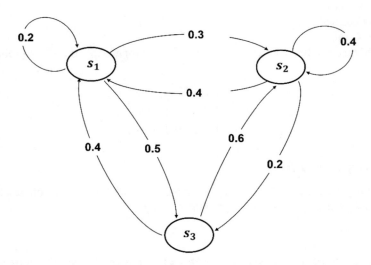

(b) The probability of transition from state s_2 to state s_3 in 2 steps:

$$p_{3/2}^{(2)} = p_{2/2} \times p_{3/2} + p_{1/2} \times p_{3/1} = 0.4 \times 0.2 + 0.4 \times 0.5 = 0.28$$

Another way to calculate this probability is by using the Chapman-Kolmogorov relations. Thus, to determine the probabilities of passing in n steps, the transition matrix at power n must be determined:

$$M^2 = \begin{bmatrix} p_{1/1}^{(2)} & p_{2/1}^{(2)} & p_{3/1}^{(2)} \\ p_{1/2}^{(2)} & p_{2/2}^{(2)} & p_{3/2}^{(2)} \\ p_{1/3}^{(2)} & p_{2/3}^{(2)} & p_{3/3}^{(2)} \end{bmatrix} = \begin{bmatrix} 0.36 & 0.48 & 0.16 \\ 0.32 & 0.40 & 0.28 \\ 0.32 & 0.36 & 0.32 \end{bmatrix}$$

resulting in $p_{3/2}^{(2)} = 0.28$

(c)   The probability of transition from state s_1 to state s_2 in 3 steps is obtained by determining the matrix M^3:

resulting in $p_{2/1}^{(3)} = 0.396$.

(d)   The state of a Markov system after a number of n steps, knowing the initial state is determined with the relation:

$$P^{(n)} = P^{(0)} * M^n$$

Thus, $P^{(3)} = P^{(0)} * M^3$

$$P^{(3)} = [0.4 \ \ 0.1 \ \ 0.5] \cdot \begin{bmatrix} 0.328 & 0.396 & 0.276 \\ 0.336 & 0.424 & 0.240 \\ 0.336 & 0.432 & 0.232 \end{bmatrix} = [0.3328 \ \ 0.4168 \ \ 0.2504]$$

(e)   The stationary state P* can be established by solving the system:

$$\begin{cases} P^* \cdot M = P^* \\ \sum_{j=1}^{M} p_j^* = 1 \end{cases} , where P^* = [p_j^*], j = \overline{1,3}$$

For our problem the system becomes:

$$\begin{cases} [p_1^* \ p_2^* \ p_3^*] \cdot \begin{bmatrix} 0.2 & 0.3 & 0.5 \\ 0.4 & 0.4 & 0.2 \\ 0.4 & 0.6 & 0 \end{bmatrix} = [p_1^* \ p_2^* \ p_3^*] \\ p_1^* + p_2^* + p_3^* = 1 \end{cases}$$

The above system leads to the following equations:

$$\begin{cases} 0.2 * p_1^* + 0.4 * p_2^* + 0.4 * p_3^* = p_1^* \\ 0.3 * p_1^* + 0.4 * p_2^* + 0.6 * p_3^* = p_2^* \\ 0.5 * p_1^* + 0.2 * p_2^* = p_3^* \\ p_1^* + p_2^* + p_3^* = 1 \end{cases}$$

By solving this system, we get the solution:

$$\begin{cases} p_1^* = 0.3333 \\ p_2^* = 0.4167 \\ p_3^* = 0.2500 \end{cases}$$

Example 6.3 [1] Find the corresponding graph for a two-step binary source (second order memory source) and calculate its entropy if $p(s_i)$ are the ones corresponding to the stationary state and the elements $p(a_j/s_i)$ are chosen randomly.

Solution Consider the binary ($M = 2$) source having the alphabet:

$$A = \{0, 1\}$$

The m = 2-step memory source will have $M^m = 4$ states:
$s_1 = 00$
$s_2 = 01$
$s_3 = 10$
$s_4 = 11$

The graph corresponding to the source can be found taking into consideration that at the emission of symbol a_j, from a state s_i only certain states can be reached:

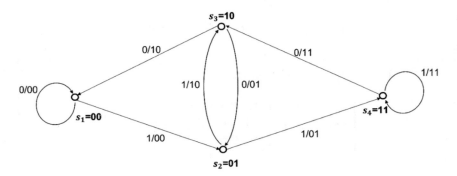

We choose the transition matrix of this source as:

$$M = \begin{bmatrix} 1/4 & 3/4 & 0 & 0 \\ 0 & 0 & 1/2 & 1/2 \\ 1/2 & 1/2 & 0 & 0 \\ 0 & 0 & 1/2 & 1/2 \end{bmatrix}$$

Notice that the matrix M contains zero elements, therefore, to calculate the stationary distribution, we should first be certain it exists, and we verify if the source is regular. For $n_0 = 2$, M^2 has all the elements strictly positive, so the source is regular, therefore it accepts a stationary state which can be established by solving the system:

$$\begin{cases} \begin{bmatrix} p_1^* & p_2^* & p_3^* & p_4^* \end{bmatrix} \cdot M = \begin{bmatrix} p_1^* & p_2^* & p_3^* & p_4^* \end{bmatrix} \\ p_1^* + p_2^* + p_3^* + p_4^* = 1 \end{cases}$$

We obtain:

$$P^* = \begin{bmatrix} \frac{2}{11} & \frac{3}{11} & \frac{3}{11} & \frac{3}{11} \end{bmatrix}$$

The elements of this source, necessary for entropy computation are contained in the table:

s_i			$p(s_i) = p$	$p(a_j/s_i)$
a_{i1}	a_{i2}	a_j		
0	0	0	2/11	1/4
0	0	1	2/11	3/4
0	1	0	3/11	1/2
0	1	1	3/11	1/2
1	0	0	3/11	1/2
1	0	1	3/11	1/2
1	1	0	3/11	1/2
1	1	1	3/11	1/2

Using the entropy calculation relation of a source with memory of m-steps:

$$H_m(A) = \sum_{i=1}^{M^m} p(s_i) H(A/s_i) = - \sum_{i=1}^{M^m} \sum_{j=1}^{M} p(s_i) p(a_j/s_i) \log_2 p(a_j/s_i)$$

$$= - \sum_{i=1}^{M^m} \sum_{j=1}^{M} p(s_{ji}) \log_2 p(a_j/s_i)$$

where $s_{ji} = \frac{a_{i1} \ldots a_{im} a_j}{s_i}$, we obtain: $H_2(A) = 0.549$ bits/binary symbol.

Example 6.4 [1] A computer game implies the use of only two keys (a_1, a_2). If the strike of a key depends on the key stroked before, a Markov chain may model the game, its transition matrix being:

$$M = \begin{bmatrix} \frac{1}{3} & \frac{2}{3} \\ \frac{3}{4} & \frac{1}{4} \end{bmatrix}$$

(a)   Which is the average quantity of information obtained at one key strike? Assume that $p(a_1)$ and $p(a_2)$ are the stationary state probabilities. Compare this value with the one obtained for a memoryless source model.

(b)   Draw the graph corresponding to the 2nd order Markov source model. How will be the average quantity of information obtained at a key strike compared to that from (a)?

Solution

(a)   The stationary state PMF can be established by solving the system:

$$\begin{cases} [p_1^*, \ p_2^*] \cdot M = [p_1^*, \ p_2^*], \\ p_1^* + p_2^* = 1 \end{cases}$$

We obtain: $P^* = \left[\frac{9}{17} \ \frac{8}{17} \right]$

The average quantity of information obtained for a key strike represents the first order Markov entropy of the source:

$$H_1(S) = - \sum_{i=1}^{2} \sum_{j=1}^{2} p(s_i) p(a_j/s_i) \log_2 p(a_j/s_i)$$

$$= - \left(\frac{9}{17} \cdot \frac{1}{3} \cdot \log_2 \frac{1}{3} + \frac{9}{17} \cdot \frac{2}{3} \cdot \log_2 \frac{2}{3} + \frac{8}{17} \cdot \frac{3}{4} \cdot \log_2 \frac{3}{4} + \frac{8}{17} \cdot \frac{1}{4} \cdot \log_2 \frac{1}{4} \right)$$

$$= 0.867 \text{ bits/symbol}$$

In the case of a memoryless source modelling of the game, we have:

$$H(S) = - \sum_{i=1}^{2} p_i \log_2 p_i = - \frac{9}{17} \cdot \log_2 \frac{9}{17} - \frac{8}{17} \cdot \log_2 \frac{8}{17} = 0,9974 \text{ bits/symbol}$$

so $H_1(S) < H(S)$, as expected.

(b)   A two-step memory binary source is modelled as described in the previous problem. The average quantity of information obtained at a key strike will be the entropy of second order Markov source and will be smaller than the one corresponding to the first order Markov source model.

Table 6.1 Create Markov chain

dtmc	Create discrete-time Markov chain
mcmix	Create random Markov chain with specified mixing structure

Table 6.2 Determine Markov Chain structure

asymptotics	Determine Markov chain asymptotics
classify	Classify Markov chain states
lazy	Adjust Markov chain state inertia
subchain	Extract Markov subchain

Table 6.3 Visualize Markov chain

distplot	Plot Markov chain redistributions
graphplot	Plot Markov chain directed graph
simplot	Plot Markov chain simulations

6.2 Applications and MATLAB Examples

6.2.1 Functions to Create, Determine and Visualize the Markov Chain Structure

To create and visualize the Markov *chan* structure, the Econometrics Toolbox from Matlab, includes the *dtmc* model object representing a finite-state, discrete-time, homogeneous Markov chain, that can be used in a variety of applications. The *dtmc* model is robust enough to serve in many modeling scenarios. We have illustrated in the Tables 6.1, 6.2 and 6.3 different functions that can be used when for the Markov chains [3].

6.2.2 Create, Modify and Visualize the Markov Chain Model

Let the transition matrix for a Markov chain:

$$M = \begin{bmatrix} 0.2 & 0.3 & 0.5 \\ 0.4 & 0.4 & 0.2 \\ 0.4 & 0.6 & 0 \end{bmatrix}$$

the Matlab code to create and visualize the Markov Chain Model is the following one:

```matlab
clear all
close all
clc

% The transition matrix for a Markov chain
M = [0.2 0.3 0.5; 0.4 0.4 0.2; 0.4 0.6 0];
% Create a Markov chain object characterized by the
transition matrix M
mc = dtmc(M)

% Display the number of states in the Markov chain.
numstates = mc.NumStates;
stateNames = ["Node1" "Node2" "Node3"];
mc.StateNames = stateNames;
% Verify that the elements within rows sum to 1 for
all rows.
sum(mc.P,2)
% Plot a directed graph of the Markov chain.
figure;
graphplot(mc);
% The transition matrix at power 2 can be deter-
mined
M_2 = M*M
% The probability of transition from state s_2 to
state s_3 in 2 steps
probability = M_2(2,3)
% The transition matrix at power 2 can be deter-
mined
M_3 = M_2*M
% The probability of transition from state s_1 to
state s_2 in 3 steps
probability = M_3(1,2)
% The state of a Markov system after a number of n
steps, knowing the initial
% state is P_0 = [0.4 0.1 0.5]
P_0 = [0.4 0.1 0.5];
P_3 = P_0 * M_3
% After creating and plotting the Markov chain, you
can determine characteristics of the chain
% The stationary state PMF is determined
PMF = asymptotics(mc)
```

In Fig. 6.1 the graph of the Markov model obtained by running the code is plotted.

Fig. 6.1 Graph of the
Markov chain

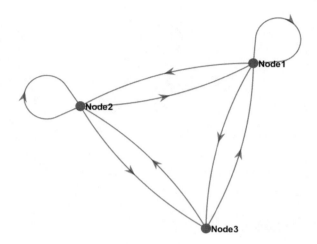

6.3 Proposed Problems [2]

Example 6.5 The simplified model of a binary transmission channel is represented
below:

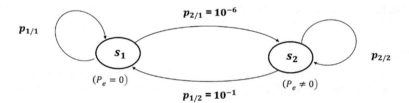

In state s_1 the channel does not introduce errors ($P_e = 0$), and in state s_2 the
transmitted binary sequences are erroneous ($P_e \neq 0$).

Determine:

(a) The probability that at the exit of the channel an error package of length $l = 5$
will appear
(b) The probabilities of the steady state
(c) The error probability P_e.

Example 6.6 Let be a Markov system characterized by the graph in the figure below:

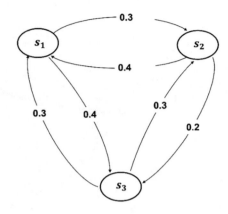

Determine:

(a)   the probability of passing from state s_3 to state s_1 in 3 steps
(b)   the state of the system after 3 steps, knowing that the initial state is $P^{(0)} =$
       $[0.2 \quad 0.3 \quad 0.5]$
(c)   the stationary state of the system.

Example 6.7 In a 4-node neural network, an impulse can only pass from one node to the consecutive node and back with a probability of 1/2. It is considered that for the last node the next node is the first, and for the first node, the previous node is the last.

   Determine:

(a)   the Markov system associated with this network (graph and transition matrix)
(b)   check if the system admits a stationary state and if so determine this state.

6.4   Conclusions

This chapter presents a series of problems solved and proposed in the field of Markov systems.

References

1. Borda M (2011) Fundamentals in information theory and coding. Springer
2. Losifescu M (1977) Lanturi Markov finite si aplicatii. Editura Tehnica Bucuresti
3. Matlab Help, www.mathworks.com

Chapter 7
Noise in Telecommunication Systems

7.1 Generalities

7.1.1 Definition and Classification

Definition The noise designates the unwanted signals that tend to disturb the transmission and processing of signals in communications systems.

Classification By taking into consideration the part of the telecommunication system where the noise appears, two types of noise can be defined: internal and external noise.

7.1.2 Internal Noise

Definition The internal noise contains all noise categories that arise inside the components of a communication system [1].

Types:

(a) *The thermal noise* is the result of the random motion of electrons in a conducting medium (resistors). The power spectrum of the thermal noise is wide and essentially uniform over the RF spectrum of interest for most communication systems.
(b) *The shot noise* arises from the discrete nature of the current flow in electronic devices (transistors, tubes). The power spectrum of the shot noise is like that of the thermal noise, and both effects are studied together.
(c) *The flicker noise (1/f noise)* occurs in active devices at low frequencies.
(d) *The quantization noise* arises in all *analog-to-digital converters*.

M. Borda et al., *Randomness and Elements of Decision Theory Applied to Signals*,
https://doi.org/10.1007/978-3-030-90314-5_7

7.1.3 External Noise

Definition The external noise contains all noise categories that arise outside the components of a communication system.

Types:

(a) *The atmospheric noise* is produced by lightning discharges during storms and its power spectrum is less than 20 MHz. Its value is maximum at the Equator, and it decreases with latitude.
(b) *The galactic noise* is caused by the disturbances originating outside the Earth's atmosphere having as a primary source the Sun. Its power spectrum ranges between 15 MHz and 500 MHz.
(c) *The man-made noise* consists of any source of electrical noise resulting from a man-made device or system (electric motors, automobile ignition systems, power lines, etc.) The noise level is higher in urban areas than in rural regions and its spectral power is less than 10 MHz.
(d) *The interferences* from other communication sources.

7.2 Thermal Noise

Definition The thermal noise is the result of the random motion of electrons in any conducting medium and it depends on the temperature.

Let us consider a resistor whose value is denoted by R. The random motion of electrons inside the resistor gives a small amount of energy that can be represented by a random value $n(t)$, illustrated in Fig. 7.1.

Properties:

1. The instantaneous amplitude of the thermal noise has a Gaussian probability density function, with a zero-mean value ($\mu_n = 0$):

$$f_N(n) = \frac{1}{\sqrt{2\pi\sigma_n^2}} e^{-\frac{n^2}{2\sigma_n^2}}.$$

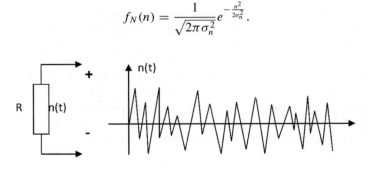

Fig. 7.1 Thermal noise generation

Fig. 7.2 Computation of the
available power of the
thermal noise

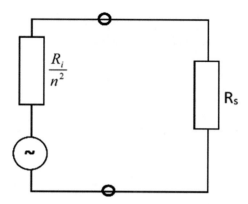

2. The mean-square noise voltage is:

$$\overline{n^2} = 4RkTB,$$

where:

- R is the resistance used to measure the thermal noise (Ω);
- k is the Boltzmann's constant $\left(1.38 \times 10^{-23}\,\text{J/K}\right)$;
- T is the temperature (K);
- B is the frequency bandwidth in which the noise is measured (Hz).

3. The thermal noise is also called white noise due to the fact that its power spectral
    density is constant for $f \in (-\infty, \infty)$.

In order to compute the power spectral density, the available thermal noise power
(P_a) is needed. Starting from the circuit given in Fig. 7.2, its value is measured on R_S:
Knowing that $R_i = R_S = R$:

$$P_a = \frac{\overline{n^2}}{4R} = kTB[W].$$

- The available power is completely independent of the resistance.

Remark

The power spectral density of the thermal noise is defined in the following for:

- One-sided representation, $f \in [0, \infty)$:

$$S_n^{(1)}(f) = \frac{P_a}{B} = kT = N_0 [W/Hz].$$

- Two-sided representation, $f \in (-\infty, \infty)$ (Fig. 7.3):

$$S_n^{(2)}(f) = \frac{N_0}{2} [W/Hz].$$

4. Assuming that the input of an ideal low-pass filter consists in white Gaussian noise, its output will also be a white Gaussian noise.
5. The *equivalent noise bandwidth* (B_n) of a transmission system is the bandwidth of a hypothetical ideal low-pass filter that produces the same output noise power as the considered transmission system when the input noise has a uniform spectral density (Fig. 7.4).

For an ideal low-pass filter, the power is given by:

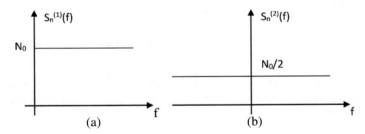

Fig. 7.3 Power spectral density of the thermal noise in **a** one-sided and **b** two-sided representation

Fig. 7.4 Ideal and real transfer function

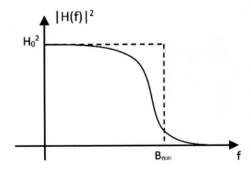

$$P_{N_{id}} = N_0 \times |H(f)|^2 \times B_n = N_0 H_0^2 B_n.$$

In real situations, the power is computed as:

$$P_{N_{real}} = N_0 \int_0^\infty |H(f)|^2 df.$$

Knowing that $P_{N_{id}} = P_{N_{real}}$, then:

$$B_n = \frac{1}{H_0^2} \int_0^\infty |H(f)|^2 df.$$

7.3 Models for the Internal Noise

7.3.1 Effective Noise Temperature

To compute the effective noise temperature T_e, the system in Fig. 7.5 is considered, knowing that:

- R is the resistor modeling the noise source;
- T_S is the source's temperature;
- T_e models the amplifier's noise;
- g is the amplifier's gain;
- B_n is the equivalent noise bandwidth.

In this case, the output noise power has two components: P'_{N_o} and P''_{N_o} due to both source noise and internal noise:

$$P_{N_o} = P'_{N_o} + P''_{N_o} = kT_S g B_n + kT_e g B_n = k(T_S + T_e)g B_n.$$

Fig. 7.5 Computation of the equivalent temperature

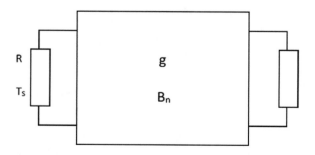

• The temperature of the noise source T_S and the effective noise temperature T_e are not physical temperatures, they are rather models. Therefore, values of the order of thousands of K are not impossible.

Remark

7.3.2   Noise Figure

The noise figure F is determined starting from the system represented in Fig. 7.6, where:

• P_{s_i} is the signal power at the input;
• P_{n_i} is the noise power at the input;
• P_{s_o} is the signal power at the output;
• P_{n_o} is the noise power at the output.

The noise factor is computed as follows:

$$F = \left. \frac{\frac{P_{s_i}}{P_{n_i}}}{\frac{P_{s_o}}{P_{n_o}}} \right|_{T_S=T_0=290K} = \frac{P_{s_i}}{P_{s_o}} \times \frac{P_{n_o}}{P_{n_i}} = \frac{1}{g} \frac{gk(T_S + T_e)B_n}{kT_S B_n}$$

$$= 1 + \left. \frac{T_e}{T_S} \right|_{T_S=T_0} = 1 + \frac{T_e}{T_0},$$

where $T_0 = 290$ K is the standard temperature.

Fig. 7.6 The computation of the noise factor

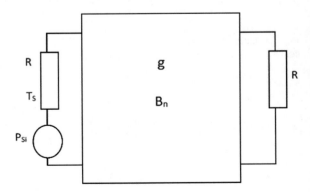

Starting from the expression of the noise figure, the effective noise temperature can be expressed as follows:

$$T_e = (F - 1)T_0.$$

- The value of the noise figure in dB can be obtained by considering the following relation:

$F[dB] = 10 \times \lg F$

- $F \in [1, \infty)$;
- $F[dB] \in [0, \infty)$.

Remark

7.4 Noise in Cascaded Systems

Let us consider the cascade of amplifiers given in Fig. 7.7 for which the equivalent effective noise temperature T_e, the equivalent gain g and the equivalent bandwidth B_n are computed.

The equivalent gain:

$$g = \prod_i g_i.$$

The equivalent bandwidth:

$$B_n = \min\{B_{n_i}\}.$$

The equivalent effective noise temperature is obtained starting from the definition relationship of the noise power at the output:

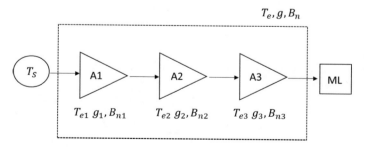

Fig. 7.7 Cascade of amplifiers

$$P_{n_o} = k(T_S + T_e)g B_n = kT_S g_1 g_2 g_3 B_n + kT_{e_1} g_1 g_2 g_3 B_n + kT_{e_2} g_2 g_3 B_n + kT_{e_3} g_3 B_n$$

$$= kg_1 g_2 g_3 \left(T_S + T_{e_1} + \frac{T_{e_2}}{g_1} + \frac{T_{e_3}}{g_1 g_2} \right) B_n = k \left[T_S + \left(T_{e_1} + \frac{T_{e_2}}{g_1} + \frac{T_{e_3}}{g_1 g_2} \right) \right] g B_n$$

$$\Rightarrow T_e = T_{e_1} + \frac{T_{e_2}}{g_1} + \frac{T_{e_3}}{g_1 g_2}.$$

- In a cascade, the input stage is the most important for the noise temperature of the system.

Remark

7.5   Effective Noise Temperature of an Attenuator

Definition A matched attenuator is a passive circuit (for which there are no other energy sources other than those produced by the thermal noise) matched at the input and output.

The transmission line is a matched attenuator.

Example

In Fig. 7.8 it is shown the model of an attenuator.
The insertion power loss:

$$L = \frac{P_i}{P_o} = \frac{1}{g} > 1.$$

Fig. 7.8 The model of an attenuator

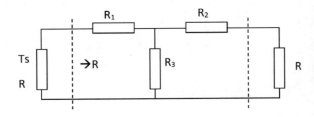

Starting from the definition of the noise power at the output, the effective noise temperature can be expressed as follows:

$$P_{n_o} = \alpha_1 k T_S B_n + \alpha_2 k T_{ph} B_n.$$

Knowing that:

$$\begin{cases} \alpha_1 + \alpha_2 = 1 \\ \alpha_1 = \frac{1}{L} = g \end{cases} \Rightarrow \alpha_2 = 1 - \frac{1}{L} \Rightarrow P_{n_o} = \frac{1}{L} k T_S B_n + \left(1 - \frac{1}{L}\right) k T_{ph} B_n$$

$$= \frac{1}{L} k [T_S + (L-1)T_{ph}] B_n \Rightarrow T_e = (L-1)T_{ph}.$$

7.6 Noise in PCM Systems

In PCM systems, the noise is due to quantization and decision [2].

7.6.1 Quantization Noise

This type of noise appears during the generation of the PCM words. For its computation, the following hypotheses are considered:

- The quantization is uniform. The *quantization step size* is determined using the relations:

$$\begin{cases} \Delta = \frac{X}{q}, \text{ if the signal x(t)is unipolar(x(t)} \in [0, X]) \\ \Delta = \frac{2X}{q}, \text{ is the signal x(t)is bipolar(x(t)} \in [-X, X]) \end{cases}$$

where $|X|$ is the maximum value of the signal x(t), q is the number of quantization levels:

$$q = 2^n,$$

and n is the length of the PCM words.

- The quantization is smooth, the probability density of the noise being uniform.

Definition The *quantization noise* n_q is defined as:

$$n_q = x_k - x_{kq},$$

where x_k represents the sample and x_{kq} represents the quantized sample.

Definition The *quantization noise power* P_{N_q} is computed using the relation:

$$P_{N_q} = \frac{1}{R}\frac{\Delta^2}{12}.$$

Definition The *signal-to-quantization noise ratio* ξ_q is given by:

$$\xi_q = \frac{P_S}{P_{N_q}} = \frac{\frac{1}{R}\overline{x^2(t)}}{\frac{1}{R}\frac{\Delta^2}{12}} = 12\frac{\overline{x^2(t)}}{\Delta^2} = 3\left(\frac{X_{ef}}{X}\right)^2 q^2.$$

For $q = 2^n$, the relation can be expressed in dB, as follows:

$$\xi_{q[dB]} = 4.7dB + 6n + 20\lg\frac{X_{ef}}{X}.$$

Remark

- The quantization noise depends on the length of the PCM word (n) and it can be reduced by increasing the number of bits.
- The quantization noise cannot be removed as it depends on the quantization process.

7.6.2 Decision Noise

This type of noise appears if the PCM words are erroneous, the quantized sequence being also erroneous. For computing the decision noise, the following hypotheses are considered:

- The considered code is the natural binary one;
- The channel is binary symmetric with independent errors;
- The length of the PCM words is n;
- In a codeword of length n, at most one error may appear;
- The error probability of a bit is p.

Definition The decision noise n_d is defined as:

$$n_d = x_{kq} - x'_{kq},$$

where x_{kq} represents the correct quantized sample and x'_{kq} is the erroneous quantized sample. If the position on which the error appears is denoted by i, the decision noise is given by the relation:

$$n_{d_i} = 2^i \times \Delta,$$

where Δ is the quantization step size.

Definition The decision noise power P_{N_d} is computed using the relation:

$$P_{N_d} = \frac{1}{R} p \frac{\Delta^2}{3} q^2,$$

where q is the number of quantization levels.

- The decision noise depends on the channel and can be reduced by improving the signal-to-noise ratio at the input of the channel.

Remark

7.6.3 The Total Signal-to-Noise Ratio in a PCM System

Definition In a PCM system, the total signal-to-noise ratio ξ_t is obtained as follows:

$$\xi_t = \frac{P_S}{P_{N_q} + P_{N_d}} = \frac{\frac{1}{R} \overline{x^2(t)}}{\frac{1}{R} \frac{\Delta^2}{12} + \frac{1}{R} p \frac{\Delta^2}{3} q^2} = \frac{12}{\Delta^2} \frac{\overline{x^2(t)}}{1 + 4pq^2} = \frac{12}{\Delta^2} \frac{X_{ef}^2}{1 + 4pq^2},$$

where P_S is the signal power.

7.7 Solved Problems

Problem 7.1 Let us consider the two amplifiers cascade from Fig. 7.9. The input is a white thermal noise $n_i(t)$ with the power spectral density in unilateral representation $N_0 = 1$ pW/Hz. It is assumed that the whole system is matched. Determine the noise power at the output of the first and second stage respectively by considering that $g_1 = 20$ dB, $g_2 = 30$ dB, $B_{n1} = 1$ MHz și $B_{n2} = 10$ kHz.

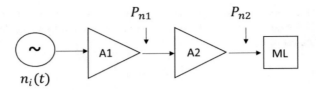

Fig. 7.9 Cascade of two amplifiers

Solution For solving the problem, the values of the gains must be linearized:

$$g_{1[dB]} = 20\,dB \Rightarrow g_1 = 10^2;$$

$$g_{2[dB]} = 30\,dB \Rightarrow g_2 = 10^3;$$

The noise power at the output of the first stage is:

$$P_{n1} = N_0 g_1 B_{n1} = 100\,\mu W.$$

The noise power at the output of the second stage is:

$$P_{n2} = \frac{P_{n1}}{B_{n1}} g_2 B_{n2} = N_0 g_1 g_2 B_{n2} = 1\,mW.$$

Problem 7.2 Let us consider a transmission system described in the block scheme from Fig. 7.10, for which $T_S = 300$ K, $T_{e_1} = 320$ K, $T_{e_2} = 420$ K, $F_3 = 3$, $g_1 = 3$ dB, $g_2 = 6$ dB, $g_3 = g_1 g_2$.

Determine the equivalent effective noise temperature of the system with respect to the antenna and the equivalent gain of the system.

Solution The equivalent effective noise temperature is given by the relation:

$$T_e = T_{e_1} + \frac{T_{e_2}}{g_1} + \frac{T_{e_3}}{g_1 g_2}.$$

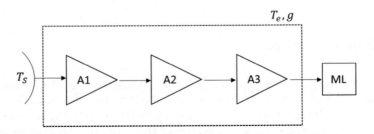

Fig. 7.10 The block scheme of a transmission system having three amplifiers

For solving the problem, the values of the gains must be linearized:

$$g_{1[dB]} = 3dB \Rightarrow g_1 = 10^{\frac{3}{10}} \cong 2;$$

$$g_{2[dB]} = 6dB \Rightarrow g_2 = 10^{\frac{6}{10}} \cong 4.$$

The value of the effective noise temperature T_{e_3} can be obtained from the definition of the noise figure:

$$T_{e_3} = (F_3 - 1)T_0 = (2 - 1) \times 290 = 290K,$$

and

$$T_e = 320 + \frac{420}{2} + \frac{290}{2 \times 4} \cong 566K.$$

The equivalent gain is the product of the gains of the three amplifiers:

$$g = g_1 g_2 g_3.$$

The values of g_1 and g_2 have been previously linearized. For g_3, the following relation is considered:

$$g_3 = g_1 g_2 = 2 \times 4 = 8.$$

The equivalent gain is:

$$g = 2 \times 4 \times 8 = 64.$$

Problem 7.3 A sinusoidal signal of amplitude $X = 1V$ is converted into a PCM signal with $\xi_q \geq 30$ dB. Determine the number of quantization levels, the length of the PCM words, and the quantization step size. Compute the total signal-to-noise ratio, knowing that the error probability of a bit is $p = 10^{-3}$.

Solution The length of a PCM word is determined starting from the definition relation of the signal-to-quantization noise ratio:

$$\xi_{q[dB]} = 4.7 + 6n + 20 \lg \frac{X_{ef}}{X}$$

$$X_{ef} = \frac{X}{\sqrt{2}}$$

$$\frac{X_{ef}}{X} = \frac{1}{\sqrt{2}}$$

$$\xi_{q[dB]} = 4.7 + 6n - 3$$

$$\xi_{q[dB]} \geq 30 \, dB \Rightarrow 1.7 + 6n \geq 30 \Rightarrow n \geq 4.72 \Rightarrow n_{min} = 5.$$

The number of quantization levels is:

$$q = 2^n = 2^5 = 32.$$

The signal is bipolar, so the quantization step size is:

$$\Delta = \frac{2X}{q} = \frac{2}{32} = 0.0625V.$$

The total signal-to-noise ratio is:

$$\xi_t = \frac{P_S}{P_{N_q} + P_{N_d}} = \frac{12}{\Delta^2} \frac{\overline{x^2(t)}}{1 + 4pq^2} = \frac{12}{\Delta^2} \frac{X_{ef}^2}{1 + 4pq^2} = 301.4.$$

7.8 Tasks

7.8.1 Problems to be Solved

Solve the following problems:

Problem 7.4 Let us consider an amplifier with the noise figure $F = 9.03$ dB, the gain $g = 50$ dB, and the bandwidth $B_n = 10$ kHz. Compute the effective noise temperature and the noise power at the output for:

(a) $T_S = T_0$;
(b) $T_S = 10 \times T_0$;
(c) $T_S = 100 \times T_0$.

Problem 7.5 Let us consider the reception system given in the Fig. 7.11, which has the bandwidth $B_n = 2$ MHz. It is assumed that the whole system is matched.

Determine:

(a)   The equivalent effective noise temperature of the reception system, with respect to the antenna.
(b)   The noise power at the output.
(c)   Using the same elements, imagine another configuration, better from the noise point of view.

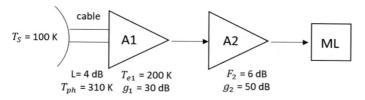

Fig. 7.11 The block scheme of a transmission system having two amplifiers

7.8.2 Practical Part in Matlab

(a) Noise in cascaded systems

Problem 7.6 Let us consider a cascade of amplifiers having the block scheme given in Fig. 7.12. Using Matlab commands, determine the equivalent gain, the equivalent effective noise temperature, and the noise figures corresponding to the considered amplifiers.

Solution

```
clc
clear all

% standard temperature
T0=290;
% enter the number of amplifiers
nr_amplifers=input('Please enter the number of ampli-
fiers in the cascade:');
g=input('Please enter the gain of all amplifiers in
the format [g1 g2 g3 ...] in dB:');
% initialization
equivalent_gain_db=0;
```

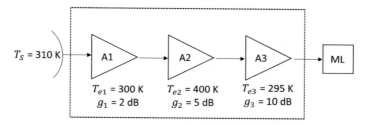

Fig. 7.12 The block scheme of the cascade of three amplifiers

```
%compute the equivalent gain both in db and linear
for i=1:nr_amplifers
    equivalent_gain_db=equivalent_gain_db+g(i);
end
equivalent_gain_linear=10^(equivalent_gain_db/10);

Te1=input ('Please enter the effective noise tempera-
ture of the 1st amplifier:');
%noise figure array
F=1+Te1/T0;
noise_temperature=0;
current_gain=1;

%consider all gains except the first one
last_gains=g(1:end-1);

%convert to linear
for i=1:(nr_amplifers-1)
    last_gains(i)=10^(last_gains(i)/10);
end

for i=1:(nr_amplifers-1)
    Te=input('Please enter the effective noise tem-
perature of the next amplifier:');
    %noise figure
    F=[F (1+Te/T0)];
    current_gain=current_gain*last_gains(i);
    current_T=Te/current_gain;
    noise_temperature=noise_temperature+current_T;
end

%equivalent effective noise temperature of the system
equiva-
lent_effective_noise_temp=Te1+noise_temperature;

%results
equivalent_gain_db
equivalent_gain_linear
equivalent_effective_noise_temp
F
```

Exercise 7.1 For the same problem (Problem 7.6), using Matlab commands, compute the power of the signal at the output of the system. Add the option to input the power of the signal emitted by the antenna. Consider the case when this power is 2 mW.

Exercise 7.2 For the same problem (Problem 7.6), using Matlab commands, compute the noise power at the output of the system. Add the option to input the noise bandwidth for the considered amplifiers. Consider the case when $B_{n1} = 2$ MHz, $B_{n2} = 5$ MHz and $B_{n3} = 3$ MHz.

Exercise 7.3 Using the Matlab code given in Problem 7.6, compute the equivalent gain, the equivalent effective noise temperature and the noise figures corresponding to the considered amplifiers given in the block scheme from Fig. 7.13.

(b) Noise in PCM systems

Problem 7.7 As input for a PCM system with n = 6 bits, a sinusoidal signal of amplitude $X = 1$ V and frequency $f = 50$ Hz is considered. Using Matlab commands, plot the input signal, the quantized one, and the PCM encoded signal. Determine the number of quantization levels, the quantization step, the signal-to-quantization noise ratio (ξ_q) and the bit rate. The sampling frequency is considered $f_s = 1$ kHz.

Solution Using the following Matlab code, we can generate a sinusoidal signal by taking as inputs the desired amplitude and frequency of the signal. The plot of the input sinusoidal wave having an amplitude $X = 1V$ and frequency $f = 50$ Hz is given in Fig. 7.14. The signal was plotted on 5 periods.

```
clc
close all
close all

frequency=input('Please enter the frequency of the
input sinusoidal signal:');
amplitude=input('Please enter the amplitude of the
input sinusoidal signal:');
```

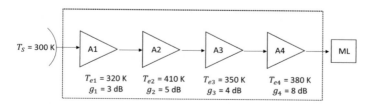

Fig. 7.13 The block scheme of the cascade of four amplifiers

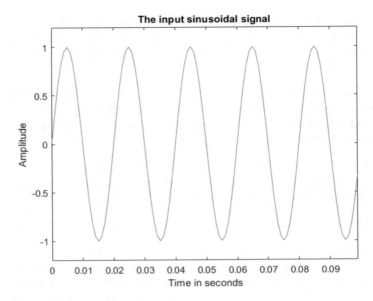

Fig. 7.14 The input sinusoidal signal

```
  sampling_frequency=input('Please enter the sampling
frequency:');
  nr_periods=input('Please enter the number of periods
used to plot the input signal:');

  %generate the sinusoidal signal
  periods=nr_periods*(1/frequency);
  t=0:1/sampling_frequency:periods-1/sampling_fre-
quency;
  w=2*pi*frequency;
  signal = amplitude*(sin(t*w));
```

The number of quantization levels, the quantization step, the signal-to-quantization noise ratio (ξ_q) and the bit rate are computed using the following Matlab code:

```
    nr_bits=input('Please enter the number of bits used
in PCM:');
    % number of quantization levels
    q=2^nr_bits;

    % bit rate
    D=nr_bits*sampling_frequency;

    % quantization step size
    total_amplitude=max(signal)-min(signal);
    delta=total_amplitude/q;

    % Quantization noise power
    Pqn=(delta^2)/12;
    Ps=sum(signal.^2)/length(signal);
    SQNR=Ps/Pqn;
    SQNR_dB=10*log10(SQNR);

    %results
    q
    D
    delta
    SQNR
    SQNR_dB
```

The quantization of the input sinusoidal signal is performed using the following Matlab code:

```
    % quantization
    scaling_factor=2/(q-1);
    scaled_signal=(signal+1)/scaling_factor;
    quantized_signal=quantiz(scaled_signal,0:q-2);
```

The PCM encoding is done by considering:

```
    % PCM encoding
    encoded_signal=[];
    for j=1:length(quantized_signal)
        encoded_signal=[encoded_signal dec2binvec(quan-
tized_signal(j),nr_bits)];
    end
```

The plots of the input signal, the quantized one and the PCM encoded signal are shown in Fig. 7.15 and are obtained using the following Matlab code:

```
figure(1);
subplot(3,1,1)
plot(t,signal);
title('The input sinusoidal signal');
ylabel('Amplitude');
xlabel('Time in seconds');
axis([-inf inf -1.2*amplitude 1.2*amplitude]);
subplot(3,1,2)
stairs(quantized_signal);
legend(sprintf('Quantized signal on %d
bits',nr_bits));
title('Quantized signal');
xlabel('Sample');
ylabel('Quantization level');
grid
subplot(3,1,3)
stairs(encoded_signal);
title('PCM encoded signal');
xlabel('Bits');
ylabel('Binary signal');
axis([0 length(encoded_signal) -1 2])
grid
```

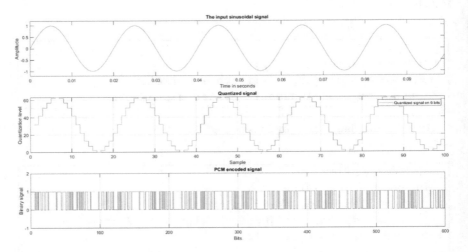

Fig. 7.15 The input sinusoidal signal, the quantized signal and the PCM encoded signal

Exercise 7.4 Using Matlab commands, compute the total signal-to-noise ratio for the signal given in Problem 7.7 by considering that the error bit probability is $p = 10^{-2}$.

Exercise 7.5 Using Matlab commands, generate and plot a sawtooth signal with a frequency $f = 50$ Hz.

Exercise 7.6 Using Matlab commands, plot the input sawtooth signal generated at Exercise 7.5, compute and plot the quantized signal and the PCM encoded signal generated from it. Determine the number of quantization levels, the quantization step, the signal-to-quantization noise ratio and the bit rate. The sampling frequency is considered $f_s = 1$ kHz.

7.9 Conclusions

The chapter introduces some theoretical aspects regarding the noise and its components and proposes some Matlab applications based on the noise theory, and also some problems and examples.

References

1. Stanley W (1982) Electronic Communications systems. Reston Pub. Co.
2. Borda M (2011) Fundamentals in information theory and coding. Springer

Chapter 8
Decision Systems in Noisy Transmission Channels

8.1 Introduction

Signal detection is part of the statistical decision theory or hypotheses testing theory. The aim of this processing, made at the receiver, is to decide which was the sent signal, based on the observation of the received signal (observation space). A block-scheme of a system using signal detection is given in Fig. 8.1.

In the signal detection block (SD), the received signal $r(t)$ (observation space) is observed and, using a decision criterion, a decision is made concerning which is the transmitted signal. The decision takes us thus the affirmation of a hypothesis. The observation of $r(t)$ can be:

- *discrete observation:* at discrete moments t_i, $i = \overline{1,N}$ samples from $r(t)$ are taken (r_i), the decision being taken on $\vec{r} = (r_1, \ldots, r_N)$. If N is variable, the detection is called sequential.
- *continuous observation:* $r(t)$ is observed continuously during the observation time T, and the decision is taken based on $\int_0^T r(t)dt$. It represents the discrete case at the limit: $N \to \infty$.

If the source S is binary, the decision is binary, otherwise M-array (when the source is M-array). We will focus only on binary detection, the M-array case being a generalization of the binary one.

The binary source is:

$$S : \begin{pmatrix} s_0(t) & s_1(t) \\ P_0 & P_1 \end{pmatrix}, \quad P_0 + P_1 = 1$$

assumed memoryless, P_0 and P_1 being the a priori probabilities.

Under the assumption of AWGN, the received signal (observation space Δ) is (Fig. 8.2):

$$H_0 : r(t) = s_0(t) + n(t) \text{ or } r/s_0$$

© The Author(s), under exclusive license to Springer Nature Switzerland AG 2021
M. Borda et al., *Randomness and Elements of Decision Theory Applied to Signals*,
https://doi.org/10.1007/978-3-030-90314-5_8

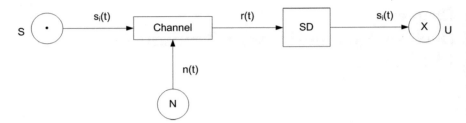

Fig. 8.1 Block scheme of a transmission system using signal detection. S–source, N–noise generator, SD–signal detection, U–user, $s_i(t)$–transmitted signal, $r(t)$–received signal, $n(t)$–noise voltage, $\hat{s}_i(t)$–estimated signal [1]

Fig. 8.2 Binary decision splits observation space Δ into two disjoint spaces Δ_0 and Δ_1 [1]

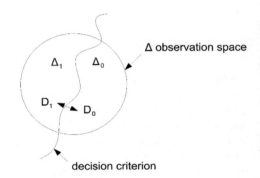

$$H_1 : r(t) = s_1(t) + n(t) \text{ or } r/s_1$$

We may have four situations:

- (s_0, D_0)—correct decision in the case of s_0
- (s_1, D_1)—correct decision in the case of s_1
- (s_0, D_1)—wrong decision in the case of s_0
- (s_1, D_0)—wrong decision in the case of s_1

The consequences of these decisions are different, and application linked; they can be valued with coefficients named costs, C_{ij}: the cost of deciding D_i when s_j was transmitted. For binary decision there are four costs that can be included in the cost matrix C:

$$C = \begin{bmatrix} C_{00} & C_{10} \\ C_{01} & C_{11} \end{bmatrix}$$

Concerning the costs, always the cost of wrong decisions is higher than those of good decisions (we pay for mistakes):

$$C_{10} \gg C_{00} \text{ and } C_{01} \gg C_{11}$$

In data transmission $C_{00} = C_{11} = 0$ and $C_{01} = C_{10}$ (the consequence of an error on '0' or on '1' is the same).

Then, for binary decision an average cost, named risk can be obtained:

$$R =: \overline{C} = \sum_{i=0}^{1} \sum_{j=0}^{1} C_{ij} P(D_i D_j) = C_{00} P(D_0/s_0) + C_{11} P(D_1/s_1) + C_{01} P(D_0/s_1)$$

$$+ C_{10} P(D_1/s_0) = C_{00} P_0 P(D_0/s_0) + C_{11} P_1 P(D_1/s_1) + C_{01} P_1 P(D_0/s_1)$$
$$+ C_{10} P_0 P(D_1/s_0)$$

Conditional probabilities $P = (D_i/s_j)$ can be calculated based on conditional pdfs (probability density functions): $p(r/s_j)$:

$$P(D_0/s_0) = \int_{\Delta_0} p(r/s_0) dr$$

$$P(D_1/s_0) = \int_{\Delta_1} p(r/s_0) dr$$

$$P(D_0/s_1) = \int_{\Delta_0} p(r/s_1) dr$$

$$P(D_1/s_1) = \int_{\Delta_1} p(r/s_1) dr$$

Taking into account that the domains Δ_0 and Δ_1 are disjoint, we have:

$$\int_{\Delta_0} p(r/s_0) dr + \int_{\Delta_1} p(r/s_0) dr = 1$$

$$\int_{\Delta_0} p(r/s_1) dr + \int_{\Delta_1} p(r/s_1) dr = 1$$

Replacing the conditional probabilities $P(D_i/s_j)$, the risk can be expressed only with one domain Δ_0, or Δ_1:

$$R = C_{11} P_1 + C_{10} P_0 + \int_{\Delta_0} [p(r/s_1)(C_{01} - C_{11}) P_1 - p(r/s_0)(C_{10} - C_{00})] dr$$

8.2   Bayes Criterion

Bayes criterion is the minimum risk criterion and is obtained minimizing (4.a):

$$\frac{p(\mathbf{r}/s_1)}{p(\mathbf{r}/s_0)} \underset{\Delta_1}{\overset{\Delta_0}{\underset{>}{<}}} \frac{P_0}{P_1} \frac{C_{10} - C_{00}}{C_{01} - C_{11}}$$

where

$$\frac{p(\mathbf{r}/s_1)}{p(\mathbf{r}/s_0)} =: \Lambda(r) =: \text{likelihood ratio}$$

$p(\mathbf{r}/s_1)$ and $p(\mathbf{r}/s_0)$ being known as likelihood functions and

$$\frac{P_0}{P_1} \frac{C_{10} - C_{00}}{C_{01} - C_{11}} = K =: \text{threshold}$$

Then Bayes criterion can be expressed as:

$$\Lambda(r) \underset{\Delta_1}{\overset{\Delta_0}{\underset{>}{<}}} K \quad \text{or} \quad \ln \Lambda(r) \underset{\Delta_1}{\overset{\Delta_0}{\underset{>}{<}}} \ln K$$

and it gives the block scheme of an optimal receiver shown in Fig. 8.3.
The quality of signal detection processing is appreciated by:

(a)   Error probability: P_E (BER)

$$P_E = P_0 P(D_1/s_0) + P_1 P(D_0/s_1)$$

Under the assumption of AWGN, the pdf of the noise is $N(0, \sigma_n^2)$:

Fig. 8.3 Block scheme of an optimal receiver (operating according to Bayes criterion, of minimum risk) [1]

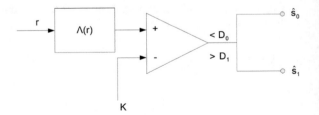

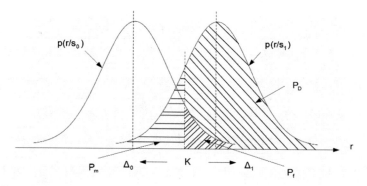

Fig. 8.4 Binary detection parameters: P_m—probability of miss, P_D—probability of detection, P_f—probability of false detection [1]

$$p(n) = \frac{1}{\sqrt{2\pi\sigma_n^2}} e^{-\frac{1}{2\sigma_n^2}n^2}$$

and the conditional pdf: $p(r/s_i)$ are also of Gaussian type (see Fig. 8.4). In engineering, the terminology, originating from radar is:

- *probability of false alarm*: P_f

$$P_f = \int_{\Delta_1} p(r/s_0)\mathrm{d}r$$

- *probability of miss*: P_m

$$P_m = \int_{\Delta_0} p(r/s_1)\mathrm{d}r$$

- *probability of detection*: P_D

$$P_D = \int_{\Delta_1} p(r/s_1)\mathrm{d}r$$

(b)   Integrals from normal *pdfs* can be calculated in several ways, one of them being the function Q(y), also called *complementary error function (co-error function: erfc).*

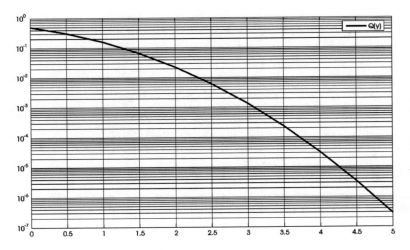

Fig. 8.5 Graphical representation of function Q(y) [1]

$$Q(y_1) =: \int_{y_1}^{\infty} f(y)\mathrm{d}y$$

where f(y) is a normal standard pdf, N(0,1):

$$f(y) =: \frac{1}{\sqrt{2\pi}} e^{-\frac{1}{2}y^2} \text{ and } \sigma^2\{y\} = 1$$

with average value $E\{y\} = y = 0$ under the assumption of ergodicity. Its graphical representation is given in Fig. 8.5.

The properties of function Q(y) are:

-

$$Q(-\infty) = 1$$

-

$$Q(+\infty) = 0$$

-

$$Q(0) = \frac{1}{2}$$

-

$$Q(-y) = 1 - Q(y)$$

If the Gaussian *pdf* is not normal standard, a variable change is used:

$$t = \frac{y - \overline{y}}{\sigma_y}, \text{ with E\{y\}} = \overline{y} \neq 0 \text{ and } \sigma_n \neq 1$$

8.3 Discrete Detection of a Unipolar Signal

Hypotheses:

- unipolar signal (in baseband): $s_0(t) = 0$

$$s_1(t) = A = \text{ct}$$

- AWGN: $N(0,\sigma_n^2)$; $r(t) = s_i(t) + n(t)$
- T = bit duration = observation time
- Discrete observation with N samples per observation time (T) $\Rightarrow \vec{r} = (r_1, \dots, r_N)$
- $C_{00} = C_{11} = 0$, $C_{01} = C_{10} = 1$, P_0, P_1, while $P_0 = P_1 = \frac{1}{2}$

a.   *Likelihood ratio calculation*

$$H_0 : \text{r}(t) = s_0(t) + n(t) = n(t) \rightarrow \text{r}/s_0 = n(t)$$

A sample $r_i = n_i \in N(0,\sigma_n^2)$ and the N samples are giving $\vec{r} = (r_1, \dots, r_N) = (n_1, \dots, n_N)$

$$p(r_i/s_0) = \frac{1}{\sqrt{2\pi\sigma_n^2}} e^{-\frac{1}{2\sigma_n^2} r_i^2}$$

$$p(\vec{r}/s_0) = \left[\frac{1}{\sqrt{2\pi\sigma_n^2}}\right]^N e^{-\frac{1}{2\sigma_n^2} \sum_{i=1}^{N} r_i^2}$$

$$H_1 : \text{r}(t) = s_1(t) + n(t) = A + n(t)$$

$$r_i = A + n_i \in N(A,\sigma_n^2) \Rightarrow p(r_i/s_1) = \frac{1}{\sqrt{2\psi_n^2}} e^{-\frac{1}{2\sigma_n^2}(r_i - A)^2}$$

$$p(\vec{r}/s_1) = \left[\frac{1}{\sqrt{2\pi\sigma_n^2}}\right]^N e^{-\frac{1}{2\sigma_n^2}\sum_{i=1}^{N}(r_i - A)^2}$$

$$\Lambda(r) = e^{\frac{A}{\sigma_n^2}\sum_{i=1}^{N} r_i} e^{-\frac{A^2 N}{2\sigma_n^2}}$$

b.   *Minimum error probability test, applied to the logarithmic relation:*

$$ln\,\Lambda(r) \underset{\Delta_1}{\overset{\Delta_0}{\lessgtr}} lnk \quad ln\,\Lambda(\vec{r}) \underset{\Delta_1}{\overset{\Delta_0}{\lessgtr}} ln\,K$$

$$\frac{A}{\sigma_n^2}\sum_{i=1}^{N} r_i - \frac{A^2 N}{2\sigma_n^2} \underset{\Delta_1}{\overset{\Delta_0}{\lessgtr}} lnk, \text{ or}$$

$$\sum_{i=1}^{N} r_i \underset{\Delta_1}{\overset{\Delta_0}{\lessgtr}} \frac{\sigma_n^2}{A} lnK + \frac{AN}{2}$$

where $\sum_{i=1}^{N} r_i$ represents a sufficient statistic, meaning that it is sufficient to take the decisions and represents a threshold depending on the power of the noise on the channel (σ_n^2), the level of the signal (A), the number of samples (N) and P_0, P_1 (through $K = P_0/P_1$).

$$\frac{\sigma_n^2}{A} lnK + \frac{AN}{2} = K'.$$

The above relation gives the block-scheme of an optimal receiver (Fig. 8.6).

If $K = 1$ and $N = 1$ (one sample per bit, taken at $T/2$) and $P_0 = P_1$ ($K = 1$), the decision relation becomes:

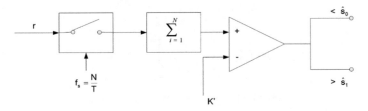

Fig. 8.6 Block-scheme of the optimal receiver for unipolar signal and discrete observation [1]

$$r_i \underset{\underset{\Delta_1}{>}}{\overset{\overset{\Delta_0}{<}}{}} \frac{A}{2}$$

c.   *Error probability of the optimal receiver variable*

According to the decision relation, the decision variable is $\sum_{i=1}^{N} r_i \in n(E[y],\sigma^2)$, $E[y]$ being the mean and σ^2 the dispersion. Making the variable change

$$y = \frac{\sum_{i=1}^{N} r_i}{\sigma_n \sqrt{N}}$$

a normalization is obtained: $\sigma^2[y] = 1$.

Using this new variable, the decision relation becomes:

$$\frac{\sum_{i=1}^{N} r_i}{\sigma_n \sqrt{N}} \underset{\underset{\Delta_1}{>}}{\overset{\overset{\Delta_0}{<}}{}} \frac{\sigma_n}{A\sqrt{N}} \ln K + \frac{A\sqrt{N}}{2\sigma_n}$$

If we note:

$$\frac{\sigma_n}{A\sqrt{N}} \ln K + \frac{A\sqrt{N}}{2\sigma_n} = \mu$$

from the above relations we get:

$$y \underset{\underset{\Delta_1}{>}}{\overset{\overset{\Delta_0}{<}}{}} \mu$$

Under the two assumptions H_0 and H_1, the pdfs of y are:

$$p(y/s_0) = \frac{1}{\sqrt{2\pi}} e^{-\frac{1}{2} y^2}$$

$$p(y/s_1) = \frac{1}{\sqrt{2\pi}} e^{-\frac{1}{2}\left(y - \frac{A\sqrt{N}}{\sigma_n}\right)^2}$$

graphically represented in Fig. 8.7.

$$P_f = \int_{\mu}^{\infty} p(y/s_0) dy = Q(\mu)$$

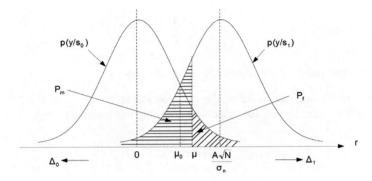

Fig. 8.7 Graphical representation of pdfs of decision variables for unipolar decision and discrete observation [1]

$$P_m = \int\limits_{-\infty}^{\mu} p(y/s_1)\mathrm{d}y = 1 - Q\left(\mu - \frac{A\sqrt{N}}{\sigma_n}\right)$$

It is a particular value μ_0 for which $P_f = P_m$:

$$Q(\mu_0) = 1 - Q\left(\mu_0 - \frac{A\sqrt{N}}{\sigma_n}\right)$$

It follows that:

$$\mu_0 = \frac{1}{2}\frac{A\sqrt{N}}{\sigma_n}$$

and μ_0 is obtained if $K = 1$, which means that $P_0 = P_1 = \frac{1}{2}$ and:

$$P_E = Q\left(\frac{1}{2}\frac{A\sqrt{N}}{\sigma_n}\right)$$

or

$$P_E = Q\left(\sqrt{\frac{1}{2}\xi N}\right)$$

where ξ designates the SNR:

$$\text{SNR} = \xi = \frac{P_s}{P_n} = \frac{\frac{A^2}{2R}}{\frac{\sigma_n^2}{R}} = \frac{1}{2}\frac{A^2}{\sigma_n^2}$$

It follows that the minimum required SNR for $P_E = 10^{-5}$, for N = 1 is approximately 15 dB (15,6 dB), which is the threshold value of the required input SNR: ξ_{i0} which separates the regions of the decision noise to that one of the quantization noise.

8.4 Discrete Detection of Polar Signal

Hypotheses:

- Polar signal in baseband: $s_0(t) = B$, $s_1(t) = A$, $B < A$, $B = -A$.

- AWGN: $N(0,\sigma_n^2)$
- T-observation time = bit duration
- Discrete observation with N samples per T
- $C_{00} = C_{11} = 0$, $C_{01} = C_{10} = 1$

Following the steps similar to those from *8.3*, we obtain:

a.

$$\Lambda(\vec{r}) = e^{-\frac{1}{2\sigma_n^2}\left[\sum_{i=1}^{N}(r_i-A)^2 - \sum_{i=1}^{N}(r_i-B)^2\right]}$$

b.

$$\ln\Lambda(\vec{r}) \underset{\underset{\Delta_1}{>}}{\overset{\overset{\Delta_0}{<}}{}} \ln K$$

$$\sum_{i=1}^{N} r_i \underset{\underset{\Delta_1}{>}}{\overset{\overset{\Delta_0}{<}}{}} \frac{\sigma_n^2}{A-B}\ln K + \frac{N(A+B)}{2}$$

For a polar signal: $B = -A$, and $K = 1$, the threshold of the comparator is $K' = 0$; if N = 1, the comparator will decide 'A' for positive samples and '0' for negative ones.

c. To calculate the quality parameters: P_E, a variable change for normalization of the decision variable is done:

$$\underbrace{\frac{\sum_{i=1}^{N} r_i}{\sigma_n \sqrt{N}}}_{y} \overset{\Delta_0}{\underset{\Delta_1}{\lessgtr}} \underbrace{\frac{\sigma_n}{(A-B)\sqrt{N}} \ln K + \frac{(A+B)\sqrt{N}}{2\sigma_n}}_{\mu}$$

The decision variable pdf under the two hypotheses is:

$$p(y/s_0) = \frac{1}{\sqrt{2\pi}} e^{-\frac{1}{2}\left(y - \frac{B\sqrt{N}}{\sigma_n}\right)^2}$$

$$p(y/s_1) = \frac{1}{\sqrt{2\pi}} e^{-\frac{1}{2}\left(y - \frac{A\sqrt{N}}{\sigma_n}\right)^2}$$

The threshold μ_0 for which $P_f = P_m$ is:

$$\mu_0 = \frac{(A+B)\sqrt{N}}{2\sigma_n}$$

which implies $K = 1$ $(P_0 = P_1 = \frac{1}{2})$.
  If $B = -A$, the polar case,

$$P_E = Q\left(\frac{A\sqrt{N}}{\sigma_n}\right) = Q\left(\sqrt{\xi N}\right)$$

Compared with the unipolar case we may notice that the same BER (P_E) is obtained in the polar case with 3 dB less SNR.

8.5 Continuous Detection of Known Signal

Hypotheses:

$$\bullet \quad \begin{cases} \qquad\quad s_0(t) = 0 \\ s_1(t) = s(t) \text{ ,of finite energy } E = \int\limits_0^T s^2(t)\mathrm{d}t \end{cases}$$

- T-observation time.
- continuous observation: $\vec{r} = (r_1, \dots, r_{N \to \infty})$
- AWGN: $n(t) \in N(0, \sigma_n^2)$, $r(t) = s_i(t) + n(t)$

a. *Calculation of* $\Lambda(r) = \frac{p(r/s_1)}{p(r/s_0)}$

Continuous observation means $N \to \infty$. We shall express the received signal r(t) as a series of orthogonal functions $v_i(t)$ in such a way that the decision could be taken using only one function (coordinate), meaning that it represents sufficient statistics.

$$r(t) = \lim_{N \to \infty} \sum_{i=1}^{N} r_i v_i(t)$$

The functions $v_i(t)$ are chosen to represent an orthonormal (orthogonal and normalized) system.

$$\int_0^T v_i(t)v_j(t)dt = \begin{cases} 1, \text{ if } i = j \\ 0, \text{ if } i \neq j \end{cases}$$

The coefficients r_i are given by:

$$r_i =: \int_0^T r(t)v_i(t)\, dt$$

and represent the coordinates of r(t) on the observation interval [0,T]. In order to have $v_1(t)$ as sufficient statistics, we chose:

$$v_1(t) = \frac{s(t)}{\sqrt{E}}$$

and r_1 is:

$$r_1 = \int_0^T r(t)v_1(t)dt = \frac{1}{\sqrt{E}} \int_0^T r(t)s(t)dt$$

We show that higher order coefficients: r_i with $i > 1$, do not affect the likelihood ration:

$$\Lambda(\vec{r}) = \frac{p(\vec{r}/s_1)}{p(\vec{r}/s_0)} = \frac{\prod_{i=1}^{N \to \infty} p(r_i/s_1)}{\prod_{i=1}^{N \to \infty} p(r_i/s_0)} = \frac{p(r_1/s_1)}{p(r_1/s_0)}$$

the contribution of higher order coefficients being equal in the likelihood ratio:

$$r_i/s_0 = \int_0^T n(t)v_i(t)dt$$

$$r_i/s_1 = \int_0^T [s(t) + n(t)]v_i(t)\mathrm{d}t = \int_0^T s(t)v_i(t)\mathrm{d}t + \int_0^T n(t)v_i(t)\mathrm{d}t$$

$$= \int_0^T n(t)v_i(t)\mathrm{d}t = r_i/s_0$$

because $\int_0^T s(t)v_i(t) = \sqrt{E}\int_0^T v_1(t)v_i(t)\mathrm{d}t = 0$, based on the orthogonality of $v_1(t)$ and $v_i(t)$.

Then, $v_1(t) = s(t)/\sqrt{E}$ is a sufficient statistic.

$$\Lambda(\vec{r}) = \frac{p(r_1/s_1)}{p(r_1/s_0)}$$

$$H_0: \mathrm{r}(t) = s_0(t) + n(t) = n(t)$$

$$r_1/s_0 = \int_0^T n(t)v_1(t)\mathrm{d}t = \frac{1}{\sqrt{E}}\int_0^T n(t)s(t)\mathrm{d}t = \gamma$$

and has a normal pdf.

The average value of $r_1/s_0 : \bar{r}_1/s_0 = 0$ based on $n(t) \in N(0,\sigma_n^2)$ and $\sigma^2[r_1/s_0] = \sigma^2[\frac{1}{\sqrt{E}}\int_0^T n(t)s(t)\mathrm{d}t] = \frac{1}{E}\sigma_n^2 TE = \sigma_n^2 T$.

It follows that:

$$p(r_1/s_0) = \frac{1}{\sqrt{2\pi\sigma_n^2 T}}e^{-\frac{1}{2\sigma_n^2 T}r_1^2}$$

$$H_1 : \mathrm{r}(t) = s_1(t) + n(t) = s(t) + n(t)$$

$$r_1/s_1 = \frac{1}{\sqrt{E}}\int_0^T [s(t) + n(t)]s(t)\mathrm{d}t$$

$$= \frac{1}{\sqrt{E}}\int_0^T s^2(t)\mathrm{d}t + \frac{1}{\sqrt{E}}\int_0^T n(t)s(t) = \sqrt{E} + \gamma$$

$$p(r_1/s_1) = \frac{1}{\sqrt{2\pi\sigma_n^2 T}}e^{-\frac{1}{2\sigma_n^2 T}\left(r_1-\sqrt{E}\right)^2}$$

$$\Lambda(\vec{r}) = e^{-\frac{1}{2\sigma_n^2 T}\left(r_1^2 - 2\sqrt{E}r_1 + E - r_1^2\right)}$$

b. *The decision criterion is:*

$$ln\Lambda(\vec{r}) \underset{\Delta_1}{\overset{\Delta_0}{\gtrless}} \ln K$$

$$r_1 \underset{\Delta_1}{\overset{\Delta_0}{\lessgtr}} \frac{\sigma_n^2 T}{\sqrt{E}}\ln K + \frac{\sqrt{E}}{2}$$

If r_1 is replaced the decision relation becomes:

$$\int_0^T r(t)s(t)dt \underset{\Delta_1}{\overset{\Delta_0}{\lessgtr}} \sigma_n^2 T\ln K + \frac{E}{2}$$

$$\sigma_n^2 T\ln K + \frac{E}{2} = K'$$

The block-scheme of the optimal receiver can be implemented in two ways: correlator-base (Fig. 8.8 top), or matched filter-based (Fig. 8.8 bottom).

c. *Decision relation. Making a variable change to obtain unitary dispersion, we get:*

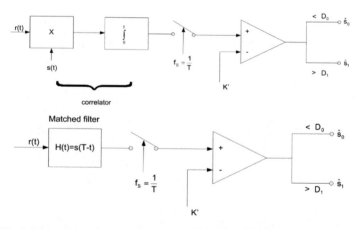

Fig. 8.8 Block Scheme of an optimal receiver with continuous observation decision for one known signal s(t): (top) correlator based implementation; (bottom) matched filter implementation [1]

$$\frac{r_1}{\sqrt{\sigma_n^2 T}} \underset{\Delta_1}{\overset{\Delta_0}{\underset{>}{<}}} \sqrt{\frac{\sigma_n^2 T}{E} \ln K} + \frac{1}{2}\sqrt{\frac{E}{\sigma_n^2 T}}$$

Using the notations:

$$z = \frac{r_1}{\sqrt{\sigma_n^2 T}}$$

$$\mu = \sqrt{\frac{\sigma_n^2 T}{E} \ln K} + \frac{1}{2}\sqrt{\frac{E}{\sigma_n^2 T}}$$

the pdfs of the new variable z are:

$$p(z/s_0) = \frac{1}{\sqrt{2\pi}} e^{-\frac{1}{2}z^2}$$

$$p(z/s_1) = \frac{1}{\sqrt{2\pi}} e^{-\frac{1}{2}\left(z - \sqrt{\frac{E}{\sigma_n^2 T}}\right)^2}$$

which are represented in Fig. 8.9.

The probabilities occurring after the decision are:

$$P(D_0/s_0) = \int_{-\infty}^{\mu} p(z/s_0)dz = 1 - Q(\mu)$$

$$P(D_1/s_1) = P_D = \int_{\mu}^{\infty} p(z/s_1)dz = Q\left(\mu - \sqrt{\frac{E}{\sigma_n^2 T}}\right)$$

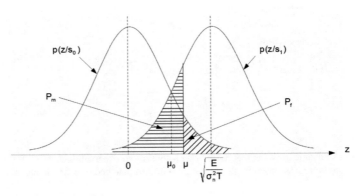

Fig. 8.9 Graphical representation of $p(z/s_0)$ and $p(z/s_1)$ [1]

$$P(D_0/s_1) = P_m = \int_{-\infty}^{\mu} p(z/s_1)dz = 1 - Q\left(\mu - \sqrt{\frac{E}{\sigma_n^2 T}}\right)$$

$$P(D_1/s_0) = P_f = \int_{\mu}^{\infty} p(z/s_0)dz = Q(\mu)$$

The particular value μ_0 for which $P_f = P_m$ is:

$$\mu_0 = \frac{1}{2}\sqrt{\frac{E}{\sigma_n^2 T}}$$

and is obtained for $K = 1$. In this case, $K = 1$, the bit error rate is:

$$P_E = Q\left(\frac{1}{2}\sqrt{\frac{E}{\sigma_n^2 T}}\right)$$

which can be expressed also as a function of the ratio E_b/N_0

$$\begin{cases} E_b = \frac{E}{2} \\ \sigma_n^2 T = N_0 BT = N_0 \frac{1}{2T} T = \frac{N_0}{2} \end{cases}$$

$$P_E = Q\left(\sqrt{\frac{E}{2} \cdot \frac{1}{2} \cdot \frac{2}{N_0}}\right) = Q\left(\sqrt{\frac{E_b}{N_0}}\right)$$

It follows that the required E_b/N_0 for 10^{-5} BER is 12.6 dB, with 3 dB less than the required ξ in the discrete observation with 1 sample per bit.

8.6 Transmission Systems with Binary Decision

We assume that the source generates only two messages: a_0 and a_1 having the probabilities P_0, respectively P_1:

$$m : \left\{ \begin{pmatrix} a_0 & a_1 \\ P_0 & P_1 \end{pmatrix} \right\}$$

By modulating these symbols, they will have in correspondence the signals $s_0(t)$, respectively $s_1(t)$.

At the receiver, two hypotheses can be made: H_0 is the hypothesis that a_0 was transmitted and H_1 is the hypothesis that a_1 was transmitted.

The reception system processes the received signal and decides which one of the two hypotheses is correct. The receiver will take the decision D_0 when hypothesis H_0 is true or it will take the decision D_1 when H_1 is considered true.

In the transmission channel the noise is added to the useful signal, so the received signal has the following form:

$$r(t) = s(t) + n(t)$$

The observation of the signal $r(t)$ can be done continuously or in certain discreet moments of time, obtaining the samples considered to be random variables. Next, it will be considered the second case.

In the observation interval T of the received signal, we will obtain the samples: $r_0, r_1, \ldots, r_N$. Those samples form a point, respectively a vector in an N-dimensional space called the observation space (interval t must be equal with the interval of a source message ai). The decision criteria allow some statements to be made (based on the observations), regarding the symbols generated by the source. The observation space can be divided into two regions: Δ_0 and Δ_1 [2].

If the representative point of the observation is formed in the region Δ_0, it can be stated that in the observation interval the received signal $r(t)$ is most likely to originate in $s_0(t)$, then in $s_1(t)$, therefor the decision D_0 will be taken. In the same way, if the representative point is formed in the region Δ_1, D_1 will be decided.

Because the transmission process there are noise interferes, it can be possible that the previous algorithm will lead to some erroneous decisions.

The most widespread decision criterion, which allows the partition of the observation space in two regions Δ_0 and Δ_1, is the Bayes criterion. For this criterion, the following must be known: the probabilities P_0 and P_1, and the costs of the possible decision (C_{ij}).

The Bayes criterion is based on the division of the space Δ in two regions Δ_0 and Δ_1, in such a way that it obtains a minimum risk. The mathematical expression of the Bayes criterion is the following one:

$$\frac{p(\vec{r}/s_1)}{p(\vec{r}/s_0)} \underset{\Delta_1}{\overset{\Delta_0}{\underset{>}{<}}} \frac{P_0(C_{10} - C_{00})}{P_1(C_{01} - C_{11})}$$

where the ratio of the density probabilities is called the likely-hood ratio (depends only on the observed data, not on the costs or the probabilities of the symbols):

$$\Lambda(\vec{r}) = \frac{p(\vec{r}/s_1)}{p(\vec{r}/s_0)}$$

The threshold of the test is:

$$K = \frac{P_0(C_{10} - C_{00})}{P_1(C_{01} - C_{11})}$$

Therefore, the Bayes rule consists in the comparison of the likely-hood ratio with the test threshold [3]:

$$\Lambda(\vec{r}) \overset{H_0}{\underset{H_1}{\lessgtr}} K$$

There is the possibility that by making a change coordinates system, all the necessary information for taking a decision to be in only one coordinate l.

The coordinate l is called sufficient statistic and allows the transition from an N-dimensional space to a space with one dimension.

In this case, the relation will be written as:

$$\frac{p(l/s_1)}{p(l/s_0)} \overset{\Delta_0}{\underset{\Delta_1}{\lessgtr}} K$$

For the two hypotheses, the received signal is:

$$H_0 : r(t) = s_0(t) + n(t)$$

$$H_1 : r(t) = s_1(t) + n(t)$$

The noise $n(t)$ is considered to be a Gaussian one of average value and the dispersion σ is known.

Applying a change of coordinates, such that all the information needed for taking a decision to be contained in only one coordinate, the Bayes criterion has the following form:

$$\int_0^T r(t)s_1(t)\mathrm{dt} - \int_0^T r(t)s_0(t)\mathrm{dt} \overset{\Delta_0}{\underset{\Delta_1}{\lessgtr}} K'$$

The integral:

$$\int_0^T r(t)s_i(t)dt = C_{rs}(0)$$

Fig. 8.10 Structure of
optimal receiver with
continuous detection of two
signals

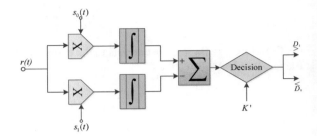

represents the mutual correlation in the origin between the received $r(t)$ and the emitted signal $s_i(t)$. It results that the Bayes decision is taken by comparing the difference between the mutual correlations in the origin with the test threshold.

For example, if $C_{rs1}(0) + K > C_s(0)$, it will be considered that the received signal in the observation interval T comes from $s_1(t)$.

The relation above suggests the next structure of the optimal receiver (see Fig. 8.10).

In the application, for the transmission of the two messages $a_0 = 0$ and $a_1 = 1$, the following signals were used:

$$s_0(t) = A \sin(\omega t)$$

$$s_1(t) = A \sin(\omega t + \pi) = -A \sin(\omega t)$$

This type of modulation is called phase shift keying. For the previous two signals we get the following relation:

$$\int_0^T r(t) sin(\omega t)\mathrm{d}t \underset{\Delta_1}{\overset{\Delta_0}{\underset{>}{<}}} K'$$

This relation generates the following structure of the optimal receiver presented in Fig. 8.11.

Fig. 8.11 Structure of
optimal receiver with
continuous detection of two
signals in phase opposition

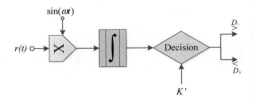

8.7 Applications and LabView Examples

8.7.1 LabView Application for the Transmission System with Binary Decision

The application for the Transmission System with Binary Decision is designed in such a way that it allows the illustration of the operation principle of a real binary decision transmission system. The application offers the user the possibility of analyzing the signals present in the operation process of a transmission system.

The front panel of the transmission system with binary decision contains all the user interaction features required to control the transmission system and it is represented in Fig. 8.12.

The front panel can be split into two main parts, the controls section, and the graphical results section.

The controls section is on the left side of the panel, and it allows the user to adjust the noise that is introduced on the transmission channel by changing the standard deviation σ_n, and thus the AWGN's amplitude either by rotating the knob or by specifying a certain value in the numeric control above the knob, see Fig. 8.13a.

The controls section allows the user to control the message that is sent from the information source, see Fig. 8.13b. There are three types of binary sequences (messages), formed from four bits, which are characterized by the probabilities of appearance of the bits:

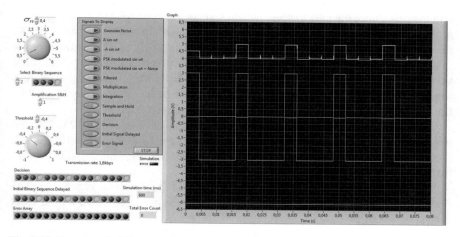

Fig. 8.12 Front panel of the transmission system with binary decision

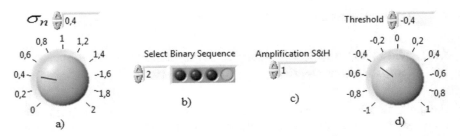

Fig. 8.13 Controls section. **a** Noise standard deviation control; **b** Binary sequence control; **c** Sample and hold amplification control; **d** Threshold control

- in the first sequence, the "0" bit has the probability $p_0 = \frac{1}{4}$ and the "1" bit has the probability $p_1 = \frac{3}{4}$;
- in the second sequence, the two bits have the same probabilities $p_0 = p_1 = \frac{1}{2}$;
- in the third sequence, the "0" bit has the probability $p_0 = \frac{3}{4}$ and the "1" bit has the probability $p_1 = \frac{1}{4}$;

The control section also allows the user to control the amplification of the sample and hold signal that is obtained after some processing that occurs at the receiver, see Fig. 8.13c.

Also, the control section has the option of adjusting the threshold value that is used in the decision process. The adjustment can be made by using the knob or specifying the desired value in the numerical control box next to the knob, see Fig. 8.13d.

The control section also contains a secondary panel that has an array of push buttons that allows the user to select which signals will be displayed on the adjacent graph. The selected signals have different offsets according to the number of signals that are selected by the user to display them simultaneously and better observe the relations between the signals. The limited space on the graph introduced the constraint that only four (five if the threshold signal is displayed over the sampled signal) signals can be displayed with offset on the graph. If more than five signals are selected by the user, they will be displayed overlapped, thus with no offset.

This secondary panel also contains a **Stop** button that allows the user to stop the simulation at any given moment and a **Simulation Error LED** that is used to identify if any errors occurred in the processing chain at the receiver. This secondary panel is shown in Fig. 8.14a.

The front panel also contains three arrays of LED, shown in Fig. 8.14b. The decision bits, initially generate bits and the error bits are also shown by the LED arrays, allowing an easier interpretation of the results. In the front panel, next to the LED arrays are two counters, one used to show the time that elapses from the beginning of the simulation, Fig. 8.14c, and the other one used to count the number of errors that were detected in the transmission system, Fig. 8.14d.

The **Decision** array represents the bits that were recovered at the receiver from the transmitted signal that was affected by the noise on the channel but in this case, there is no noise.

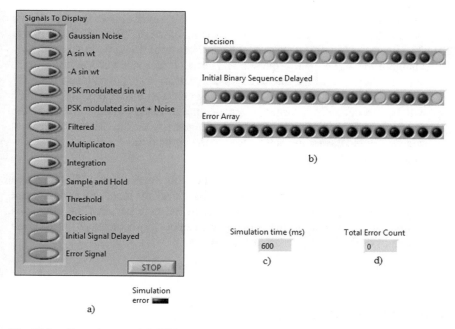

Fig. 8.14 **a** Secondary panel; **b** LED arrays; **c** simulation time; **d** error counter

The **Initial Binary Sequence** array contains the bits that were transmitted at the source but delayed being able to compare them with the decision array of bits obtained at the receiver so that we can determine if errors occurred during the transmission process. These errors are shown in the **Error** array on the position they occurred.

The significance of the LEDs is as follows. If the LED is ON it means that the bit recovered/ sent is a "1" bit and if it is OFF then the bit is a "0". In the case of the Error array, if the LED is ON then an error has occurred at the receiver on that specific position in the array, and if the LED is OFF then the bit was successfully transmitted from the source to the receiver.

The last component of the front panel is the Graph indicator which displays the signals that are selected in the secondary panel. The graph can be seen on the right side of Fig. 8.12, to the right of the secondary panel that controls the display.

8.7.2 Tasks

For the LabView application the following tasks are proposed:

1. Open LabView and browse for the Decision Main.vi
2. On the front panel apply the sequence 1010 at the input. Set the noise at the minimum level, the S&H Amplification at 1 and the threshold level at 0. Observe

the operation of the system and visualize all the waveforms of the signals one by one or in groups for a better understanding of the processes that take place.

3. Observe the effect of the noise on the waveforms for the same input sequence.

4. Maintain the same input sequence and set a certain value of the noise level. Observe the effect of the threshold on the decision waveform and on the error waveform. It can be noticed, that depending on the threshold level, the symbols will be erroneous in a larger or smaller number. The optimal value of the threshold is the one corresponding to the situation when the number of errors is minim. Try increasing the noise and find the optimal threshold, Explain.

5. Calculate the bit error rate:

   – maintain at the input the sequence 1010, choose a certain level of the standard deviation for the noise (between 2 and 5) and set the threshold at its optimal value. The bit error rate will be equal to the ratio between the erroneous bits and the number of transmitted bits. Repeat the experiment for different values of the noise standard deviation.

6. Repeat points 4 and 5 for the sequences 1110 and 0001.

8.7.3 Problems Proposed

Problem 1. Let the transmission system with the noise matrix $P(Y/X) = \begin{bmatrix} 0.8 & 0.1 & 0.1 \\ 0.2 & 0.2 & 0.6 \end{bmatrix}$. At the input of the channel, the symbols x_0 and x_1 were transmitted with probabilities $p_0 = 2/3$ and $p_1 = 1/3$. Knowing that the cost matrix is $C = \begin{bmatrix} 0.1 & 0.9 \\ 0.8 & 0.4 \end{bmatrix}$, determine:

(a) The partition of the received symbols using the Bayes criterion
(b) The strategy matrix S
(c) The noise matrix of the equivalent channel
(d) The average error before and after the partition
(e) The error probability and the correct probability.

Problem 2. A transmission system that uses discrete detection at the receiver with N samples per bit ($p_0 = p_1 = 1/2$, A = 6 V). The noise is AWGN with $\sigma_n = 3$ V.

(a) Which is the binary sequence at the output of the detection block if the sequence of samples at the receiver is 5,2 V; 2,5 V; 1,5 V; 2,8 V; 4,2 V; 1,9 V and N = 1? What if N = 3?
(b) Determine the minimum SNR so that $P_e \leq 10^{-5}$
(c) Determine the minimum number of samples per bit at the receiver so that $P_e \leq 3 \cdot 10^{-4}$.

Problem 3. A transmission system that uses discrete detection at the receiver with N samples per bit ($p_0 = p_1 = 1/2$, $B = -A = -6$ V). The noise is AWGN with $\sigma_n = 3$ V.

(a) Which is the binary sequence at the output of the detection block if the sequence of samples at the receiver is 0,9 V; $-1,2$ V; 0,6 V; $-0,6$ V; 1,2 V; $-0,8$ V ($N = 2$)?

(b) Determine the error probability if $N = 4$.

(c) Determine the minimum number of samples per bit at the receiver so that $P_e \leq 3 \cdot 10^{-5}$.

Problem 4. In a transmission system that uses continuous detection at the receiver, the signals $s_0(t) = 0$ and $s_1(t) = A \cdot \sin\omega_p t$ were transmitted. For s_1, $f_p = 500$ Hz and the probabilities are $p_0 = p_1 = 1/2$. The signal observation period is $T = 16$ ms, and the noise is AWGN with $\sigma_n^2 = 9$. ($P_f = P_m$).

Determine:

(a) The amplitude of the signal s_1, A, so that the error probability is

$$P_e \leq 7 \cdot 10^{-3}$$

(b) The optimal threshold at the receiver.

8.8 Conclusions

The chapter presents a LabVIEW application that allows simulations of real transmissions systems for transmission systems with a binary decision. The LabVIEW environment offers very useful tools in analyzing the transmission systems along with the operations and processes that take place at both the transmitter and receiver sides. The timing and duration of these operations and processes are important to be taken into consideration for the application, to offer accurate simulations of the transmission system. The theoretical aspects introduce the chapter.

References

1. Borda M (2011) Fundamentals in information theory and coding. Springer
2. Spataru Al (1971) Teoria transmisiunii informaţiei, vol. II. Ed. Tehnica
3. Radu M, Stoica S (1988) Telefonie Numerică. Ed. Militară, Bucureşti

Chapter 9
Comparison of ICA Algorithms for Oligonucleotide Microarray Data

9.1 Introduction

ICA (Independent Component Analysis) is a method in which the purpose is to find a linear representation of nongaussian data so that the components are statistically independent (or as independent as possible). The fact that it looks for components that are both statistically independent and non-gaussian distinguishes ICA from other methods. This technique can be applied in various circumstances, including for feature extraction or signal separation [1].

ICA is directly connected to the BSS (Blind Source Separation) method. BSS implies the separation of a set of signals from a set of mixed signals without having information about the source signals or the mixing process.

9.2 Blind Source Separation

The classical example of source separation is the "cocktail party problem" where several people are talking simultaneously in a room and one person tries to follow one of the speakers. It might be easy for a human being to handle this kind of source separation, but it is a very delicate problem in signal processing.

Suppose that in a room there are four persons speaking at the same time and four microphones that record four signals: $x_1(t), x_2(t), x_3(t), x_4(t)$. The recorded signals can be written as a weighted sum of the unobservable utterances produced by the speakers: $s_1(t), s_2(t), s_3(t), s_4(t)$. If we put these pieces of information into equations, we obtain the following:

$$x_1(t) = a_{11}s_1 + a_{12}s_2 + a_{13}s_3 + a_{14}s_4$$
$$x_2(t) = a_{21}s_1 + a_{22}s_2 + a_{23}s_3 + a_{24}s_4$$
$$x_3(t) = a_{31}s_1 + a_{32}s_2 + a_{33}s_3 + a_{34}s_4$$

© The Author(s), under exclusive license to Springer Nature Switzerland AG 2021
M. Borda et al., *Randomness and Elements of Decision Theory Applied to Signals*,
https://doi.org/10.1007/978-3-030-90314-5_9

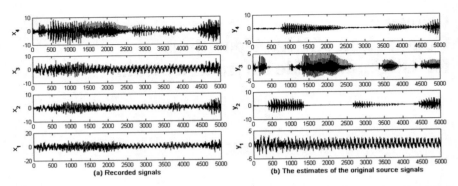

Fig. 9.1 a The signals corresponding to four speakers recorded by the microphone; **b** The original sources signals after ICA was applied [2]

$$x_4(t) = a_{41}s_1 + a_{42}s_2 + a_{43}s_3 + a_{44}s_4$$

The $a_{ij}; i, j = 1 \div 4$ coefficients are parameters that depend on the microphone to speaker distance. The idea is to try and estimate the four original signals $s_1(t), s_2(t), s_3(t), s_4(t)$ from the recorded signals $x_1(t), x_2(t), x_3(t), x_4(t)$. This would be easy if the parameters a_{ij} would not be unknown.

Let's take as an example the recorded waveforms from Fig. 9.1. They represent the signals taken from the four microphones placed in front of the speakers.

We can use some pieces of information on the statistical properties of the signals $s_i(t)$ to estimate the a_{ij} parameters we need to try and estimate the original signals from the recorded signals. Two assumptions are made. One supposes that the original signals $s_1(t), s_2(t), s_3(t), s_4(t)$ are at each moment in time t is made and the other one states that the mixing coefficient matrix $[a_{ij}]$ is invertible. We can further assume that there exists a matrix W with w_{ij} coefficients so that it can be used to separate the original signals $s_i(t)$:

$$s_1(t) = w_{11}x_1 + w_{12}x_2 + w_{13}x_3 + w_{14}x_4$$
$$s_2(t) = w_{21}x_1 + w_{22}x_2 + w_{23}x_3 + w_{24}x_4$$
$$s_3(t) = w_{31}x_1 + w_{32}x_2 + w_{33}x_3 + w_{34}x_4$$
$$s_4(t) = w_{41}x_1 + w_{42}x_2 + w_{43}x_3 + w_{44}x_4$$

We can estimate the W matrix as the inverse of the matrix formed by the mixing coefficients a_{ij} taking into consideration the assumptions that were made. Therefore, to estimate the w_{ij} coefficients statistical properties are used. If the signals are not Gaussian, it is enough to determine the w_{ij} coefficients so that the following signals are statistically independent [2]:

$$y_1(t) = w_{11}x_1 + w_{12}x_2 + w_{13}x_3 + w_{14}x_4$$
$$y_2(t) = w_{21}x_1 + w_{22}x_2 + w_{23}x_3 + w_{24}x_4$$

$$y_3(t) = w_{31}x_1 + w_{32}x_2 + w_{33}x_3 + w_{34}x_4$$
$$y_4(t) = w_{41}x_1 + w_{42}x_2 + w_{43}x_3 + w_{44}x_4$$

If the signals $y_1(t)$, $y_2(t)$, $y_3(t)$, $y_4(t)$ are statistically independent they can be considered accurate estimates of the original signals $s_i(t)$, the only difference between the two being some scalar factors. Taking into account the independence conditions mentioned, four original sources signals were separated from their mixture (Fig. 9.1).

9.3 Theoretical Aspects of ICA

To define ICA we can use a statistical "latent variables" model, assuming that we observe n linear mixed signals $x_1, x_2, x_3, \ldots, x_n$ of n independent components.

$$x_j = a_{j1}s_1 + a_{j2}s_2 + \ldots + a_{jn}s_n \; for \; all \; j$$

In the Independent Component Analysis, we suppose that each mixed signal x_j together with the independent component s_k is a random variable instead of a periodic signal.

It's suitable to use vector–matrix notation instead of the sums from the above equation. By x it is denoted the random vector whose elements are the mixtures $x_1, \ldots, x_n$, by s the random vector with $s_1, \ldots, s_n$ elements, and by A denote the matrix with a_{ij} elements.

Using vector–matrix notation the ICA mixing model can be written as:

$$x = As$$

In some cases, we need the columns a_j of the A matrix, therefore the above model can be written as:

$$x = \sum_{i=1}^{n} a_i s_i$$

The ICA model is a generative model, which means that it describes how the observed data is generated by mixing the s_i components. As previously mentioned, independent components are latent variables which means that they cannot be directly observed. We must estimate both A and s from the only thing that we can observe - the random vector x, the mixing matrix is assumed to be unknown.

The starting point for ICA is the assumption that the s_i components are statistically independent. We must also assume that the independent components must have nongaussian distributions. For simplicity reasons we also assume that the unknown mixing matrix is square. After estimating the A matrix, we can calculate its inverse $W = A^{-1}$ and obtain the independent component by:

$$s = Wx$$

9.4 ICA Algorithms

9.4.1 FastICA Algorithm

In practice, we need an algorithm to maximize the contrast function. FastICA introduces a very efficient method of maximization that would suit this task. The data must be preprocessed by centering and whitening.

Centering is the most basic preprocessing with the purpose to center x, which implies subtracting its mean vector m = E{x} so that x will become a zero-mean variable. This pre-processing is made to simplify the ICA algorithms.

Another useful preprocessing is the whitening of the observed variables. Before applying the ICA algorithm, as soon as centering has been performed, the observed vector x is linearly transformed so that we obtain a vector $\tilde{x}$ which is white. This means that its components are uncorrelated and their variances equal unity.

FastICA for one unit

FastICA for one unit refers to a computational unit, eventually an artificial neuron with the weight vector w that can be updated by a learning rule. The FastICA learning rule finds a direction through a unit vector w so that the projection $w^T x$ maximizes non-gaussianity. The variance of $w^T x$ must be constrained to unity.

FastICA is based on a fixed-point iteration scheme for finding a maximum of non-gaussianinty for $w^T x$. It can also be derived as an approximate Newton iteration if we denote by g the derivative of a nonquadratic function G.

The basic form of the FastICA algorithm is the following:

1. *Choose an initial weight vector* w
2. *Let* $w^+ = E\{x g(w^T x)\} - E\{g'(w^T x)\}w$
3. *Let* $w = w^+ / \|w^+\|$
4. *If not convergent, go back to 2.*

Convergence means that the old and new values of w point in the same direction which means that their dot-product (scalar product) is almost equal to one.

FastICA for several units

The previous one-unit algorithm estimates just one of the independent components. To estimate several independent components, we need to run the one-unit FastICA algorithm using several units with weight vectors $w_1, \ldots, w_n$.

To prevent different vectors from converging to the same maximum we must decorrelate the $w_1^T x, \ldots, w_n^T x$ outputs after every iteration. There are three methods of achieving this decorrelation.

The first method implies that we estimate the independent component one by one. After estimating p independent components or p vectors $w_1, \ldots, w_p$ we run the one-unit fixed-point algorithm for w_{p+1} and after every iteration step subtract from w_{p+1} the projections $w_{p+1}^T w_j w_j$, $j = 1, \ldots, p$ of the previously estimated p vectors, and then renormalize w_{p+1} by:

1. $Let\ w_{p+1} = w_{p+1} - \sum_{j=1}^{p} w_{p+1}^T w_j w_j$
2. $Let\ w_{p+1} = w_{p+1} / \sqrt{w_{p+1}^T w_{p+1}}$

Sometimes it may be desired to use a symmetric decorrelation where no vectors are privileged over others. We can accomplish this by the classical method involving matrix square roots [1]:

$Let\ (W = (WW)^{T-1/2} W,\ W = (w_1, \ldots, w_n)^T$

Properties of the FastICA Algorithm

1. The convergence is cubic, under the assumption of the ICA data model. Ordinary ICA algorithms have linear convergence. This means that the convergence of FastICA is very fast – confirmed by simulations and experiments.
2. No step size parameters are needed, contrary to gradient-based algorithms – the algorithm is easy to use.
3. The algorithm finds directly independent components of any nongaussian distribution using any nonlinearity g.
4. The independent components can be estimated one by one, decreasing the computational load of the method in cases when only some independent components must be estimated.
5. The FastICA algorithm has most of the advantages of neural algorithms: it is parallel, distributed, computationally simple and requires small memory space. The classic stochastic gradient methods are preferable only if fast adaptivity in a changing environment is required [3].

9.4.2 JADE Algorithm

Joint Approximate Diagonalization of Eigen-matrices (JADE) ICA approach is based on the diagonalization of cumulant matrices. For simplicity, only the case of symmetric distributions is considered so that order cumulants vanish.

For random variables $X_1, X_2, X_3, X_4, X_i^* \stackrel{\text{def}}{=} X_i - E(X_i)$, the second order cumulants denoted as $C(X_1, X_2)$ are:

$$C(X_1, X_2) \stackrel{\text{def}}{=} E\left(X_1^* X_2^*\right)$$

and the fourth-order cumulants denoted as $C(X_1, X_2, X_3, X_4)$ are:

$$C(X_1, X_2, X_3, X_4) \overset{\text{def}}{=} E(X_1^* X_2^* X_3^* X_4^*) - E(X_1^* X_2^*)E(X_3^* X_4^*) - E(X_1^* X_3^*)E(X_2^* X_4^*)$$
$$- E(X_1^* X_4^*)E(X_2^* X_3^*)$$

The variance and the kurtosis of random variable X are defined as follows:

$$\sigma^2(X) \overset{\text{def}}{=} C(X, X) = E(X^{*2}),$$
$$kurt(X) \overset{\text{def}}{=} C(X, X, X, X) = E(X^{*4}) - 3E^2(X^{*2})$$

Under a linear transformation $Y = AX$, the cumulants of the fourth-order transformation are:

$$C(Y_i, Y_j, Y_k, Y_l) = \sum_{pqrs} a_{ip} a_{jq} a_{kr} a_{ls} C(X_p, X_q, X_r, X_s)$$

Since the ICA model is linear, we use the assumption of independence by:

$$C(S_p, S_q, S_r, S_s) = kurt(S_p)\delta_{pqrs}$$

where δ is the following function:

$$\delta_{pqrs} = \begin{cases} 1 \text{ if } p = q = r = s \\ 0 \text{ otherwise} \end{cases}$$

We obtain the cumulants of $X = AS$ in which S has independent entries,

$$C(X_i, X_j, X_k, X_l) = \sum_{m=1}^{n} kurt(S_m) a_{im} a_{jm} a_{km} a_{lm}$$

where a_{ij} is the i-th row and j-th column entry of matrix A.

The JADE algorithm is specifically a statistic-based technique. Its greatest advantage is that it works off-the-shelf (requires no parameter tuning) [4]. It can be summarized as:

1. *Initialization. Estimate a whitening matrix $\widehat{W}$ and set $Z = \widehat{W}X$.*
2. *Form statistics. Estimate a maximal set $\left\{ \widehat{Q}_i^Z \right\}$ and cumulant matrices.*
3. *Optimize an orthogonal contrast. Find the rotation matrix V such that the cumulant matrices are as diagonal as possible i.e. solve $\widehat{V} = argmin \sum_i Off\left(V^T \widehat{Q}_i^Z V\right)$*
4. *Separate. Estimate connectivity matrix A by $\widehat{A} = \widehat{V}\widehat{W}^{-1}$. Then we can recover the sources S by $\widehat{S} = \widehat{A}^{-1}X$.*

9.4.3 EGLD Algorithm

The second algorithm studied in this paper employs Extended Generalized Lambda Distribution (EGLD) for modeling source distributions. The major benefit EGLD is that it takes into account the skewness of the distributions.

This algorithm proposes a source distribution adaptive approach to maximum likelihood estimation of ICA model. The EGLD modeling approach provides a useful connection between the practical estimator and theoretical measure of independence. As mentioned above, the model covers an extensive range of skewness and kurtosis values that characterize a wide class of distributions of interest in engineering and data analysis.

The lambda distribution purpose is in fitting a distribution to the empirical data, and in the computer generation of different distributions. The latest extension of the family is a combination of Generalized Lambda Distributions (GLD) and Generalized Beta Distribution (GBD).

The Extended Generalized Lambda distribution is a large family of distributions. It is defined by the inverse distribution function:

$$F^{-1}(p) = \lambda_1 + \frac{p^{\lambda_3} - (1-p)^{\lambda_4}}{\lambda_2},$$
$$f(x) = C\beta_2^{-(\beta_3 + \beta_4 + 1)}(x - \beta_1)^{\beta_3}(\beta_1 + \beta_2 - x)^{\beta_4}$$

where $0 \leq p \leq 1$ and $\lambda_1, \lambda_2, \lambda_3, \lambda_4, \lambda_3, \lambda_4$ are the parameters of the distribution.

Observations are easily generated from EGLD using the inverse distribution function above.

ICA using the EGLD Model

The underlying source distributions are estimated through the marginal distributions by fitting them to EGLD family using the method of moments. Taking into consideration that the density function of the EGLD is not available in a closed form, the score function will be obtained by deriving the inverse distribution function.

If we consider $p = F(y)$, where $F(y)$ represents the distribution function of a GLD, the score function is the following:

$$\varphi(p) = -\frac{\lambda_2 p^{\lambda_3 - 2}(\lambda_3 - 1)\lambda_3}{\left(p^{\lambda_3 - 1}\lambda_3 + (1-p)^{\lambda_4 - 1}\lambda_4\right)^2} + \frac{\lambda_2(1-p)^{\lambda_4 - 2}(\lambda_4 - 1)\lambda_4}{\left(p^{\lambda_3 - 1}\lambda_3 + (1-p)^{\lambda_4 - 1}\lambda_4\right)^2}$$

In some algorithms, the derivative of the score function is also needed.

The actual algorithm optimizing the derived criterion could be any suitable ICA algorithm where maximum likelihood contrasts are utilized. In our experiments we used natural gradient or relative gradient algorithm:

$$W_{k+1} = W_k + \eta\left(I - \varphi(y)y^T\right)W_k$$

where η represents the learning rate and fixed-point algorithm

$$W_{k+1} = W_k + D\big(E\{\varphi(y)y^T\} - diag(E\{\varphi(y_i)y_i\})\big)W_k$$

where $D = diag(\frac{1}{E\{\varphi(y_i)y_i\}} - E\{\varphi'(y_i)y_i\})$.

A 3 step algorithm is proposed for the EGLD-ICA [5].

Repeat until convergence

1. *Calculate the third and fourth sample moments α_3 and α_4 for current data $y_k = W_k x$ and select the GLD if $\alpha_4 > 2.2 + 2 * \alpha_3{}^2$ and else the GBD.*
2. *Estimate parameters for GLD or GBD by the method of moments and calculate score function $\varphi(y_k)$.*
3. *Calculate the demixing matrix W_{k+1} using the relative gradient algorithm and the fixed point algorithm.*

9.4.4 Gene Expression by MAS 5.0 Algorithm

The expression level of the data is measured using the MAS 5.0 (MicroArray Suite). MAS 5.0 is based on the Tukey-Biweight algorithm, whereas its predecessor – MAS 4.0 – is based on a weighted average calculation of the probe-pairs differences. Alongside the Tukey-Biweight algorithm, Discrimination Score and IM (Ideal Mismatch) calculation is employed. MAS 5.0 transforms the intensities to a logarithmic scale before computing the average – therefore the contribution of different probes is equalized.

Steps of the MAS 5.0 algorithm:

1. *The discrimination score (R)* – is a relative measure of the difference between Perfect Match and Mismatch intensities. The discrimination score is compared to a threshold $t = 0.0015$ and the probes which do not satisfy the $R < t$ condition are not considered. It is computed by:

$$R = \frac{PM_i - MM_i}{PM_i + MM_i}$$

2. *One step Tukey-Biweight algorithm*

The whole Tukey-Biweight algorithm performs iterations through a process of computing the estimate and analyzing until no further changes are detected. The first step of the biweight provides the most useful increase in the quality, therefore only this one-step part was considered [6].

The Specific Background (SB) is computed for the entire probeset using the equation:

$$SB = Tukey\ Biweight\big(\log_2(PM_j) - \log_2(MM_j)\big), \quad j = \overline{1, n}$$

3.   *IM (Ideal Mismatch)*

There are three cases of determining the IM for a probe pair j:

$$IM = \begin{cases} MM, & if \ MM < PM \\ \frac{PM}{2^{SB}}, & if \ MM \geq PM \ and \ SB > 0.03 \\ \frac{PM}{2^{(0.03/[1+(0.03-SB)/10])}}, & if \ MM \geq PM \ and \ SB \leq 0.03 \end{cases}$$

The 0.03 value represents the default contrast and the 10 value represents the cutoff that describes the variability of the probe pairs in the probe set. The first case when $IM = MM$ is preferred because the MM value provides a probe-specific estimate. The second case estimate is not probe-specific but provides information specific to the probe set. The third case performs the least informative estimated, very weakly based on probe-specific data.

4.   *Expression level*

The gene expression level is calculated using the TukeyBiweight algorithm:

$$Signal = TukeyBiweight\left(\log_2\left(PM_j - MM_j\right)\right), \quad j = \overline{1, n}$$

The Tukey-Biweight algorithm can be summarized below:

a.   Median calculation – the median of the data is necessary to obtain the center

$$M = median(x)$$

b.   Distance from the median – the distance of each data point from the median is needed

$$Dist(i) = abs(M - x(i))$$

c.   Median of obtained distance calculation

$$M' = median(Dist)$$

d.   A uniform measure of distance and weights for each point
e.   Finding out the result

9.5 Robust Estimation of Microarray Hybridization

The Affymetrix technology started in the later 80s with the revolutionary theory of uniting the semiconductor manufacturing techniques with the progress of combinatorial chemistry to build large sizes of biological data on a small glass chip [6].

The oligonucleotide arrays are specially manufactured chips that extract from the gene's mRNA sequences of 25 nucleotides which best represent the gene (probes).

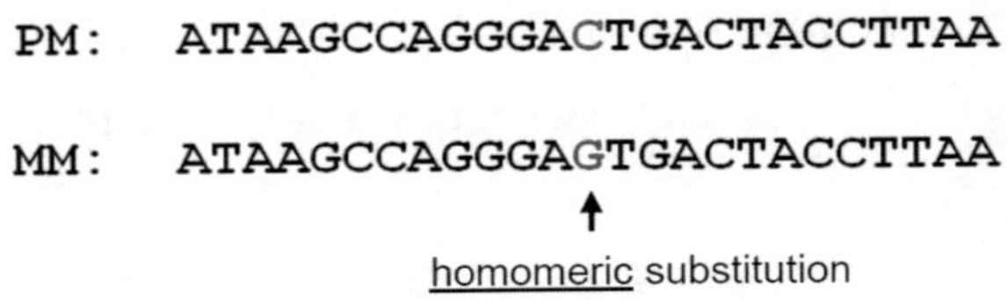

Fig. 9.2 PM and MM   PM :   ATAAGCCAGGGACTGACTACCTTAA
sequences [6]

MM :   ATAAGCCAGGGAGTGACTACCTTAA
                                              ↑
                                      homomeric substitution

The probes can have from 12 to 25 pairs of nucleotides. For these extracted sequences two pairs of vectors of the same length are synthesized—one representing the perfect complementary sequence to the extracted sequence (PM – PerfectMatch) and the other representing a complementary sequence but with the 13th position, nucleotide copied directly from the extracted sequence, creating a mismatch (MM).

An example of 25 probe pairs of PM and MM sequences is shown in Fig. 9.2.

The amounts of hybridized material for the PM—perfect match $(x_{i,k}^P)$ and MM—mismatch $(x_{i,k}^m)$ corresponding to gene i with k probe pairs can be expressed as:

$$x_{i,k}^P = \rho(s_{i,k}, x, y) p_t(s_{i,k}|z_{i,k}^P)$$
$$x_{i,k}^m = \rho(s_{i,k}, x, y) p_t(s_{i,k}|z_{i,k}^m)$$

where $\rho(s_{i,k}, x, y)$ is the surface distribution (density) of segment k in gene i, for the point from (x,y) and $p_t(s_{i,k}|z_{i,k}^P)$ and $p_t(s_{i,k}|z_{i,k}^m)$ are the hybridization probabilities for the segment $s_{i,k}$ on the test segments $z_{i,k}^P$ (for PM) and $z_{i,k}^m$ (for MM). These probabilities should satisfy the relation:

$$0 \leq p_t\left(s_{i,k}|z_{i,k}^{m/p}\right) \leq 1, \ \forall i, k; t \geq 0$$

which means that the hybridization probability between a segment i and a probe k ranges between 0 and 1 for any time interval different than zero.

In most of the cases the hybridization probabilities for PM test segments are expected to be larger than for MM ones, assuming that the hybridization process is time invariant and that the thermodynamic conditions under which hybridization is carried out are the same:

$$p_t(s_{i,k}|z_{i,k}^P) \geq p_t(s_{i,k}|z_{i,k}^m); \forall t$$

It can be hypothesized that hybridization progresses such that the probability will follow up a process in time which can be expressed as:

$$p_t\left(s_{i,k}|z_{i,k}^{p/m}\right) = 1 - e^{-t/\tau_{i,k}^{p/m}}; \forall t$$

$\tau_{i,k}^{p/m} > 0$ is the time constant for a given PM or MM pair of segments i, k, these time constants being related to the hybridization temperature of each hybridizing segment. It is predicted that the probability of hybridization will tend to zero if not enough time interval is provided for the experiment to take place, and that if more than enough limit is allowed then saturation will take place.

For short time intervals the hybridization probabilities can be considered linear:

$$p_t(s_{i,k}|z_{i,k}^{p/m}) \approx t/\tau_{i,k}^{p/m}; t \ll \tau_{i,k}^{p/m}$$

In this situation both probabilities can be considered as proportional:

$$p_t(s_{i,k}|z_{i,k}^{p}) = \eta_{i,k}p_t(s_{i,k}|z_{i,k}^{m}); t \ll \min\{\tau_{i,k}^{p}, \tau_{i,k}^{m}\}$$

with the resulting proportional parameter:

$$\eta_{i,k} = \frac{p_t(s_{i,k}|z_{i,k}^{p})}{p_t(s_{i,k}|z_{i,k}^{m})} = \frac{\tau_{i,k}^{p}}{\tau_{i,k}^{m}}$$

The proportionality parameter can be used to determine homogeneity, considering it for the test segment k independently of the gene segments:

$$\eta_{i,k} = \eta_{j,k}; \forall i, j$$

9.6 Reasons for Inconsistent Test Pair Hybridization Patterns

For a given gene i, it may not be expected that all the time constants of its associated test segment pairs $\{\tau_{i,k}^{p,m}\}$ are to be the same. Most of the time this will not happen, therefore the relative ratios between the slopes of PM and MM pairs will be different, and segments that have reached saturations will be found, whereas others will still be in the linear side of the hybridization.

When the process of hybridization for the MM pair is faster than the process for the PM pair, a saturation unbalance will appear in the alignment between perfect match and mismatch pairs. Therefore, each Probe Set can be considered as composed of two vectors: $x_i^{p} = \{x_{i,k}^{p}\}$ and $x_i^{m} = \{x_{i,k}^{m}\}$, corresponding to the perfect and mismatch for a given test.

In microarray processing it is expecting that the proportionality between the conditioned hybridization probabilities for the mismatch and perfect match is maintained: $\eta_{i,k} = \eta_{j,k} = \eta_k$ and in such cases, the expression results for the k segment could be reliable (reliably expressed cases). In some of the cases, there are some perfect match pairs with proportionality that does not match that of others within the same Probe Set (unreliably expressed cases).

To measure if a probe set is reliably or unreliably expressed the inner product between PM and MM vectors can be used:

$$\langle x_i^p x_i^m \rangle = \rho^2(s_{i,k}) \sum_{k=1}^{K} p_t(s_{i,k}|z_{i,k}^p) p_t(s_{i,k}|z_{i,k}^m)$$

and consequently, if the norm of the PM is evaluated: [7]

$$\|x_i^p\|^2 = \rho^2(s_{i,k}) \sum_{k=1}^{K} p_t^2(s_{i,k}|z_{i,k}^p)$$

The proportionality parameter for gene k given by λ_i could be expressed as:

$$\lambda_i = \frac{\|x_i^m\| \cos \beta_i}{\|x_i^p\|} = \frac{\langle x_i^m, x_i^p \rangle}{\|x_i^p\|} = \frac{\sum_{k=1}^{K} p_t(s_{i,k}|z_{i,k}^p) p_t(s_{i,k}|z_{i,k}^m)}{\sum_{k=1}^{K} p_t^2(s_{i,k}|z_{i,k}^p)}$$

We can observe from the last equation that the proportionality parameter—λ_i is strictly independent of the density distribution, therefore it is not affected by the overall gene expression. The proportionality parameter is directly related to the projection of the vectors of conditioned probabilities for the PM and MM test probes, and therefore to the hybridization time constants.

It was assumed that a single pair would not be reliable enough therefore k different test segments were used in the same probe set. The first algorithms for evaluating the gene expression were designed to perform an average on the differential expression between perfect match and mismatch. If the mismatch probe would express much more than the perfect match a problem on how to handle this data appears. This case will not respect the homogenous equation. Several causes may be possible: background noise, failures in the masking process while microarray is prepared, hybridization crosstalk between segments or other underlying processes separating the hybridization process from the ideal conditions.

To deal with the non-homogenous case and its consequences on the perfect match – mismatch alignment we consider the following graph. In Fig. 9.3, x_i^p and x_i^m are the PM and MM test vectors and x_i^c and x_i^o are the collinear and orthogonal components of x_i^m with respect to x_i^p and β_i is the angle between the mismatch and perfect-match vectors.

If the homogeneity condition would be true, a high degree of collinearity between the PM and MM vectors must be measured, with the orthogonal component being almost negligible. If the orthogonal component is significant, one can conclude that the estimated expression levels are the product of underlying random or noisy processes. Orthogonality can be measured by the following coefficient:

$$\gamma_i = 1 - cos^2 \beta_i$$

Fig. 9.3 The geometrical
relation between PM-MM
vectors

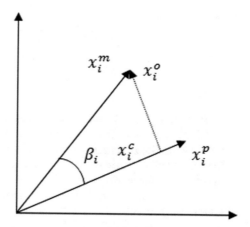

γ_i ranges between 0 for collinear vectors and 1 for orthogonal ones. Once this model is established powerful projection methods could be used to pre-process the microarray information prior to component separation and classification.

Samples with large λ_i and small γ_i may be relied upon as good estimates and their total expression can be evaluated from a complementary function of hybridization amounts. If the values of γ_i are closer to 1 this will mean that the two vectors are not correlated, a case in which independent components can be estimated by using ICA, and then derived for further research.

9.7 Reliably and Unreliably Expressed Test Probes

The robust estimation methodology for microarray hybridization has been applied on microarray data from several databases:

- **HG-U133 A**—is part of the GeneChip® Human Genome category of chips. This class of chips provides the broadest coverage of the human genome in different array formats. The most important feature of the GeneChips is that every array provides multiple, independent measurements for each transcript. Multiple probes mean obtaining a complete data set with reliable results from every experiment.

  The GeneChip® Human Genome U133 Set provides two array coverage of the transcribed genome. It allows the analysis of over 33,000 genes.

  GenBank®, dbEST, and RefSeq supplied the sequences which were used in the design of the array. The clusters were created from the UniGene database (Build 133, April 20, 2001) and then polished by comparison with several other publicly available databases.

  In situ synthetization is used to transpose on the array the oligonucleotide probes which are complementary to each other [6].

Table 9.1 γ thresholds

$\gamma \geq 0.5$	Very unreliable probes
$0.1 \leq \gamma < 0.5$	Unreliable probes
$0.05 \leq \gamma \leq 0.1$	Reliable probes
$\gamma < 0.05$	Very reliable probes

In this chapter, 2 tumors from Human Genome U133 Set A databases provided by CNIO were used. Set A contains 22283 probe sets and 10 QC (Quality Control) units, while Set B contains 22645 probe sets and 10 QC units.

- **MG-U74Av2**—provides extensive coverage of the mouse genome. The user has the ability of measuring expression levels for more than 36,000 mouse genes. The Murine Genome U74v2 Set contains the broadest transcript coverage of mouse genes. They are used to analyze mouse cells and tissues. The set is composed of three arrays and contains 36,000 full-length genes and EST (Expressed Sequence Tags) clusters. The represented sequences are derived from the clusters in Build 74 on the UniGene Database. The clusters are represented by one or more sequences derived directly from cluster members [6]. In this chapter the MG_U74Av2 chip was used, containing 12,488 probe sets and 13 QC (Quality Control) units.
- **cvyMORF**—a public database containing 7 experiments from mouse tissues on MOR chips with 1824 probe sets and 4 QC q-units. For this data, the γ parameter was computed and 3 thresholds were established: 0.05, 0.1, 0.5 (Table 9.1).

From the above table—the probe sets with values of γ larger than 0.5, the probes are considered very unreliable, and the probe sets with γ values between [0.1; 0.5] are considered unreliable. These are the values of γ that we will attempt to correct by the 5 presented ICA algorithms. The γ values between 0.05 and 0.1 and smaller than 0.05 are reliable, and respectively very reliable. These values of γ will not be corrected.

The databases consist of four.CDF files (HG-U133A.CDF and HG-U133B.CDF for the Human Genome U133, MG_U74Av2.CDF for the Murine Genome U74v2 and cvyMORF.CDF for the OE database). The Chip Description File (CDF) is provided by Affymetrix and represents the layout of the chip, describing which probes are part of which probe set. All probes set names within an array are unique. Multiple copies of a probe set may exist on a single array if each copy has a unique name.

The data collected from the Affymetrix scanner is stored into DAT files which will contain all the pixel intensity values collected. The CEL file stores the results of the intensity calculations on the pixel values of the DAT file. This includes an intensity value, standard deviation of the intensity, the number of pixels used to calculate the intensity value and others.

This data flow can be summarized in Fig. 9.4.

For each CDF file seven different CEL files were used in the processing. An initial processing was done for each CEL file, storing the necessary information into.mat

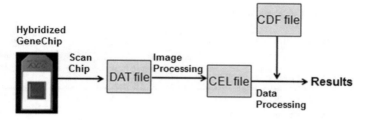

Fig. 9.4   The Affymetrix data flow [8]

files. The necessary information means the number of genes, the values of γ (the PM-MM orthogonality measure) for the very unreliable/unreliable/reliable/very reliable conditions, a structure containing the PM and MM values for each gene, the before processing gene expression computed by means of a MAS5 algorithm, and other important factors.

In the next step, ICA algorithms were applied to the CEL files and then the important data was saved in.mat files which will be later used for the algorithms' performance evaluation. The stored data will contain the values of γ after correction (for the unreliable and very unreliable case), the PM-MM values after processing for each gene or the resulting gene expression.

9.8   Applications and Matlab Implementation

The Matlab code of this implementation is listed in Appendix B, and in the further paragraphs only some results and comments are indicated.

9.8.1   Collecting the Initial Data

The databases are organized into 3 groups corresponding to each chip (CDF): the cvyMORF group (6 OE CEL files), the MG_U74Av2 group (7 CEL files) and the HG-U133A group (2 CEL files). For each group the value of the orthogonality parameter—γ was collected and structured into tables regarding the reliability of the gene. The details related to the first group of 6 CEL files can be examined in Table 9.2.

It can be observed that most of the genes have the values for the orthogonality parameter which do not fulfill the hypothesis of co-linearity and hence the reliability thresholds—about 80% of the genes are grouped in unreliable and very unreliable categories. For these categories the ICA algorithms will be applied. The data corresponding to the second group of 7 CEL files is structured in Table 9.3.

Table 9.2 The number of genes for the cvyMORF database and the organization according to the γ thresholds

Experiment name	Genes no	Very reliable ($\gamma < 0.05$) (%)	Reliable ($0.05 \leq \gamma \leq 0.1$) (%)	Unreliable ($0.1 \leq \gamma < 0.5$) (%)	Very unreliable ($\gamma \geq 0.5$) (%)
OE_2mo_F0	1824	7.35	11.90	72.59	8.17
OE_2mo_F1	1824	7.84	12.66	71.22	8.28
OE_2mo_F2	1824	8.22	12.17	73.03	6.58
OE_2mo_M0	1824	7.51	10.58	72.48	9.43
OE_2mo_M1	1824	8.66	11.07	71.93	8.33
OE_2mo_M2	1824	7.24	11.46	72.59	8.72

Table 9.3 The number of genes for the MG_U74Av2 database and the organization according to the γ thresholds

Experiment name	Genes no	Very reliable ($\gamma < 0.05$) (%)	Reliable ($0.05 \leq \gamma \leq 0.1$) (%)	Unreliable ($0.1 \leq \gamma < 0.5$) (%)	Very unreliable ($\gamma \geq 0.5$) (%)
1 h-Tc	12,488	18.86	26.02	54.36	0.77
1 h-Tc-control	12,488	14.23	22.76	62.07	0.94
2 h-Tc	12,488	17.79	27.01	54.56	0.65
2 h-Tc-control	12,488	14.52	21.40	61.58	2.50
4 h-Tc	12,488	36.23	29.49	34.10	0.28
4 h-Tc-control	12,488	31.55	28.73	39.32	0.40
48 h-Tc	12,488	45.35	28.05	26.41	0.19
48 h-Tc-control	12,488	30.73	28.92	39.99	0.37

In this case from the total number of genes—12,488—the unreliability stands between about 26 and 64% of the genes, which indicates an improvement compared to the first group of CELs. The largest databases contain 22,283 genes and because of the large amount of computation time necessary only two of them were studied. The unreliability of these experiments is like the one of the first group of CEL files, ranging around 74%. The number of unreliable genes is quite large therefore a correction is necessary. The distribution of the genes can be analyzed in Table 9.4.

Table 9.4 The number of genes for the HG-U133A database and the organization according to the γ thresholds

Experiment name	Genes no	Very reliable (γ < 0.05) (%)	Reliable (0.05 ≤ γ ≤ 0.1) (%)	Unreliable (0.1 < γ < 0.5) (%)	Very unreliable (γ ≥ 0.5) (%)
Tumores_A6T (20)HG-U133A	22,283	11.58	14.39	67.47	6.56
Tumores_A16T2 (C)HG-U133A	22,283	10.06	14.99	70.07	4.88

9.8.2 Data After ICA Processing

Let us consider the cvyMORF database. After applying the ICA Algorithms improvements in co-linearity are present. The original unreliable and very reliable stats appear on the first line of Table 9.5, and on the next lines the effect of each algorithm is noted.

For the FastICA algorithm from the total of the average 72.30% (740 genes) of unreliable data for the OE CEL files an average of 27.81% (205) were corrected – which translates into a computed γ ≤ 0.1. For the average 8.51% (87 genes) of the very unreliable data, 66.89% (58 genes) were corrected. For the EGLD algorithm from the total of 72.30% of unreliable data of the cvyMORF chip, an average of 27% of the genes were corrected; from 8.51% of the very unreliable data an average of 66.89% were corrected. The performances are almost like FastICA.

The JADE algorithm has corrected 25.35% from the average of 72.30% of unreliable data and only 22.48% of the very unreliable data. It can be easily observed that the performance of this algorithm in the case of very unreliable data is lower than the one of the previous three algorithms: FastICA and EGLD.

In Fig. 9.5 the results for an unreliable probe set which after EGLD computing became reliable (change for γ from 0.945 to 0.085) can be examined.

9.9 Conclusions

The Chap. 9 presents three Independent Components Algorithms that can be applied on microarray data: FastICA, EGLD and JADE. The microarray technology is a very useful tool in analyzing the gene expression and therefore it became widely acknowledged in the research for different diseases and the behavior of healthy and treated cells. The conditions under which the microarray chips are manufactured and then utilized are extremely important, temperature or time influences on the reliability of the data must be taken into consideration. ICA helps to better understand and manage the microarray databases in conditions of noise. By applying ICA to three

Table 9.5 The number of corrected probes for the cvyMORF chip when ICA is used

Experiment Algorithm		OE_2mo_F0	OE_2mo_F1	OE_2mo_F2	OE_2mo_M0	OE_2mo_M1	OE_2mo_M2
Without ICA	Unrel [%]	72.59	71.22	73.03	72.48	71.93	72.59
	Vunrel [%]	8.17	8.28	6.58	9.43	8.33	8.72
FastICA	Unrel [%]	28.32	29.95	27.25	26.32	28.58	26.51
	Vunrel[%]	66.44	72.19	64.17	62.21	69.08	67.30
EGLD	Unrel [%]	27.57	26.40	26.80	27.23	26.37	27.64
	Vunrel[%]	68.46	70.20	71.67	74.42	80.26	76.73
JADE	Unrel [%]	25.68	26.25	25.75	24.58	24.85	25.00
	Vunrel [%]	17.45	23.84	19.17	26.74	26.32	21.38

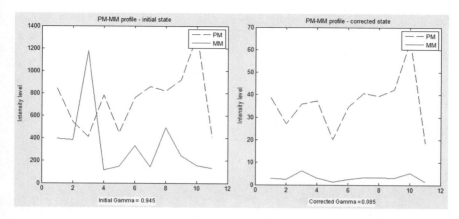

Fig. 9.5 Results for a very unreliable probe set. An γ value improvement, after EGLD computing, from 0.945 to 0.085 can be observed

different gene chips the quality of the data was notably improved hence creating a better reliability of the further processing or analysis.

References

1. Hyvärinen A, Oja E (2000) Independent component analysis: algorithms and applications. Neural Netw 13(4–5):411–430
2. Malutan R (2010) New methods in microarray statistical processing for genomic research from independent component analysis and superior order statistics. PhD Thesis, Technical University of Cluj Napoca
3. Hyvärinen A (1999) Fast and robust fixed-point algorithms for independent component analysis. IEEE Trans Neur Netw 10(3):626–634
4. Cardoso JF (1999) High-order contrasts for independent component analysis. Dissertation Ecole Nationale Superieure des Telecommunications, Paris
5. Eriksson J, Karvanen J, Koivunen V (2000) Source distribution adaptive maximum likelihood estimation of ICA model. Dissertation, Signal processing Laboratory - Helsinki University of Technology
6. Affymetrix – Statistical Algorithms Description Document (2002) http://www.affymetrix.com/. Accessed 1 September 2021
7. Malutan R, Gomez P, Borda M, Diaz F (2007) Estimation reliability of microarray data using high order statistics. Acta Tehnica Napocensis. 48(4):23–26
8. Stemcore Laboratories (2006) Affymetrix microarrays, stem cell network, microarray course. http://www.ogic.ca/projects/SCNcourse/course_units/unit1/lecture/Introduction%20to%20A ffymetrix%20Microarrays.ppt. Accessed 15 June 2020

Chapter 10
Image Classification Based on Statistical Modeling of Textural Information

10.1 Texture in Image Processing

Texture represents an important aspect in visual perception, involved in the characterization and identification of the objects around us.

In recent years, this property has been extensively studied in image analysis and several databases containing different texture samples have been created. One example is the VisTex database [1], illustrated in Fig. 10.1.

As seen in the previous figure, adjectives like "fine", "coarse", "rough", "smooth", "regular", "irregular", "metallic", "wooden", etc. are frequently used to describe textures. Being a complex element, the texture is a valuable information in computer vision used for image segmentation, image retrieval, or classification.

In this work, the textural information is used for image classification.

10.2 Texture Classification

Even though textures are easily identified and classified by human beings, they are very hard to translate into some mathematical models. One strategy to obtain relevant features is the multiscale image decomposition.

In this context, the image classification workflow consists of two steps: feature extraction and classification. First, during the feature extraction stage, the texture is analyzed using the multiscale representation. Next, the obtained coefficients are modeled by a probability density function with a specific parameter vector. By estimating these parameters, the signature of the image is obtained. In the end, during the classification stage, a similarity measure based on a probabilistic metric is computed between the signature vectors [2].

© The Author(s), under exclusive license to Springer Nature Switzerland AG 2021
M. Borda et al., *Randomness and Elements of Decision Theory Applied to Signals*,
https://doi.org/10.1007/978-3-030-90314-5_10

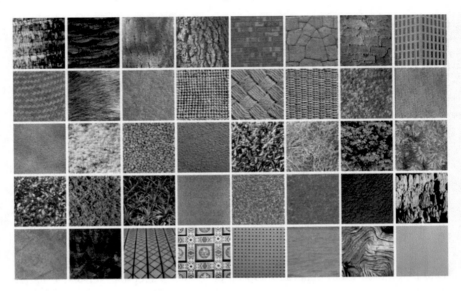

Fig. 10.1 VisTex texture database [1]

10.2.1 Feature Extraction

Multiscale approaches have been developed based on the study of human visual perception. The research carried out in this direction has shown that the human brain is able to perform a multiscale analysis of images [3, 4]. In this context, the Fourier transform, the Gabor filters, the wavelet transform, the curvelets etc. can be used for capturing the textural information.

Figure 10.2 shows an example of texture (sand), analyzed at two different resolutions. In the first image, the sand is analyzed from afar. Therefore, the texture is given by the "waves" the wind has created on the beach. In the second image, the

Fig. 10.2 Texture analyzed at two different scales

sand is closely analyzed, and the texture is given by the sand crystals. Depending on the chosen scale, the information obtained by texture analysis is different. To capture this type of information, the wavelet decomposition has been chosen.

Wavelet decomposition has been introduced in [5] and it represents an approach for multiresolution image processing. By using this technique, the image is decomposed in orthogonal and independent subbands, obtained by considering some basis functions.

In practice, for image decomposition, filter banks of low pass filters (L) and high pass filters (H) are applied along the rows and columns. As a result, four subbands are obtained by combining the two filters. These subbands consist of the image approximation (LL), horizontal (LH), vertical (HL) and diagonal (HH) coefficients. The decomposition is a recursive process and for the next level, the LL subband is used, as it can be seen in Fig. 10.3. In addition, a downsampling by a factor of two is considered at each level.

To obtain this type of image decomposition, discrete wavelet functions, such as Haar and Daubechies wavelets can be employed.

In the end, the coefficients in each wavelet subband can be described by using statistical models [2].

Once extracted, the wavelet coefficients are statistically modeled to obtain the final texture signature. Among all the possible distributions, the zero-mean multivariate Gaussian distribution has been chosen.

Multivariate Gaussian distributions with zero mean can be defined as follows.

Let $X = \{x_1, \ldots, x_N\}$ be a set of N independent and identically distributed random vectors of dimension m, issued from a zero-mean multivariate Gaussian distribution. The probability density function describing the set is the following:

$$f_X(x) = \frac{1}{\sqrt{(2\pi)^m |M|}} \exp\left(-\frac{1}{2} x^T M^{-1} x\right),$$

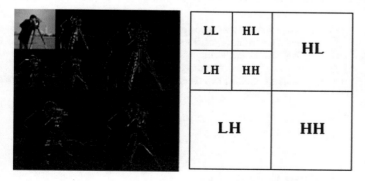

Fig. 10.3 Image wavelet transform with two levels and three orientations

where M is the covariance matrix. In the context of image classification based on wavelet coefficients, this matrix must be estimated for each wavelet subband.

Sample covariance matrix, or the empirical covariance matrix, is one of the most common estimators for covariance matrices, as it represents the solution of the maximum likelihood estimator for zero-mean Gaussian distribution. In this case, the SCM estimator $\widehat{M}$ of the covariance matrix M, characterizing X, is given by the following equation:

$\widehat{M} = \frac{1}{N}\sum_{i=1}^{N} x_i x_i^T$, where $(\cdot)^T$ denotes the transpose operator.

The set of all estimated covariance matrices represent the final textural descriptor, also called feature vectors, and they are used further in classification.

10.2.2 Classification Based on the Kullback–Leibler Divergence

During the classification process, the textures are separated into groups, or classes, knowing that the elements in the same class have similar properties. To measure the degree of similarity between two images, a distance, or a divergence is considered.

For this work, the textures are modeled by covariance matrices. In this case, one of the most common similarity measures is represented by the Kullback–Leibler divergence. For two matrices $\widehat{M}_1$ and $\widehat{M}_2$ the divergence is given by:

$$KL\left(\widehat{M}_1, \widehat{M}_2\right) = \frac{1}{2}\left[\text{tr}\left(\widehat{M}_2^{-1}\widehat{M}_1\right) - m - \ln\frac{\left|\widehat{M}_1\right|}{\left|\widehat{M}_2\right|}\right],$$

where $\text{tr}(\cdot)$ is the trace operator and m is the dimension of the vector space.

- When wavelet decomposition is used, the feature vectors contain several covariance matrices (one for each wavelet subband). Statistically, the decomposition subbands are independent, so the total distance between two images is given by the sum of all Kullback–Leibler divergences computed between pairs of subbands.

Remark

10.3 Matlab Implementation

In this section, some of the theoretical elements will be implemented.

10.3.1 Wavelet Decomposition

As seen earlier, the first step of the classification algorithm implies the extraction of the wavelet coefficients. For this application, the Daubechies 4 wavelet transform is considered. To obtain the wavelet coefficients, the Matlab functions wavedec2, detcoef2 and appcoef2 are used. The code for extracting the level one detail and approximation coefficients for an image is denoted by I_1 is the following:

```
% Image decomposition by using db4 wavelet transform
[c,s]=wavedec2(I1,2,'db4');
% Extraction of level one detail coefficients
[H1,V1,D1] = detcoef2('all',c,s,1);
% Extraction of level one approximation coefficients
A1 = appcoef2(c,s,'db4',1);
```

- To display the image corresponding to a specific subband, the *wcodemat* can be employed. For instance, the image obtained for the horizontal coefficient will be given by:
  H1_img = wcodemat(H1,255, 'mat',1);
  imshow(uint8(H1_img));

Remark

10.3.2 Covariance Matrix Estimation

For each wavelet subbands (horizontal, vertical, and diagonal), the corresponding covariance matrix has to be estimated. At this stage, to capture the spatial information, a $w \times w$ neighborhood is extracted for each pixel in the selected subband. Then, the neighbourhood's elements are stacked to form a vector. For instance, if $w = 3$ a vector with 9 elements is obtained, as shown in Fig. 10.4. The set of all vectors is then modeled by a zero-mean multivariate Gaussian distribution. The parameter of this distribution is the covariance matrix, estimated by the SCM algorithm. In the end, by concatenating all the covariance matrices, the feature vector characterizing the image textural information is obtained.

The following code can be used to estimate the covariance matrix of the horizontal subband:

Fig. 10.4 Extraction of the spatial information

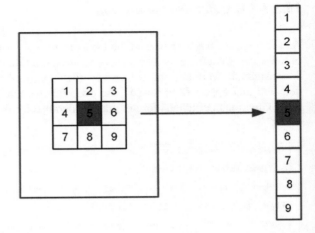

```
% Neighbourhood dimension

w = 3;


% Subband size

[m,n]=size(H1);


% Auxiliar variable

c = 0;


% Extraction of all neighbourhoods and their concatenation to form a
vector

for i = 1+floor(w/2):m-floor(w/2)

    for j = 1+floor(w/2):n-floor(w/2)

        c = c+1;

        temp = H1(i-floor(w/2):i+floor(w/2),j-
floor(w/2):j+floor(w/2) );

        x(:,c)=temp(:);

    end

end


% Covariance matrix estimation based on the extracted wavelet
coefficients

N = size(x,2);

M1 = (x*x')/N;
```

10.3.3   Kullback–Leibler Divergence

In the classification part, the similarity between the two images I_1 and I_2 has to be computed. More precisely, the Kullback–Leibler divergence between the corresponding covariance matrices is needed.

For two covariance matrices M_1 and M_2, the Kullback–Leibler divergence can be implemented as it follows:

```
% Implementation of the Kullback-Leibler divergence
if (det(M1)==0)|| (det(M2)==0)
    KL = Inf;
else

    KL = 1/2 * (trace(pinv(M2)*M1) - log(det(M1)/det(M2)) -
size(M1,1));
End
```

- In practice, the symmetric version of the Kullback–Leibler divergence, called the Jeffreys divergence, can be considered:

$$d = \tfrac{1}{2} \times [KL(M_1, M_2) + KL(M_2, M_1)]$$

where $KL(\cdot)$ is the Kullback–Leibler divergence

Remark

10.4   Tasks

Make a document containing the solutions for the following exercises:

1. Create a Matlab function that performs the Daubechies 4 wavelet transform of an image, considering 2 decomposition scales.
2. Create a Matlab function to estimate the covariance matrix for all the extracted subbands from one image.
3. Create a Matlab function to compute the total Kullback–Leibler divergence between two wavelet decomposed images.
4. Let consider two images I_1 and I_2 belonging to two different texture classes C_1 and C_2. A new image, I_3 has to be classified in one of the two classes, by using the method presented in this chapter. Integrate the previously developed code to implement the classification algorithm, knowing that I_3 will be part of the class corresponding to the minimal distance.
5. Test other types of wavelet filters. Are there any differences concerning the results?

6.   Choose a texture image database (VisTex, Brodatz, etc.) and implement a classification algorithm based on the method presented in this chapter. Propose a method to compute the classification accuracy.

10.5   Conclusions

This chapter introduces an application of statistical modeling for texture image classification. It contains both theoretical and Matlab implementation for texture characterization by wavelet decomposition, coefficients modeling by means of zero-mean multivariate Gaussian distributions along with the Sample Covariance Matrix estimation procedure, and Kullback–Leibler divergence-based classification.

References

1. Vision Texture Database. MIT Vision and Modeling Group. http://vismod.medis.mit.edu/pub/VisTex
2. Do MN, Vetterli M (2002) Wavelet-based texture retrieval using generalized Gaussian density and Kullback-Leibler distance. IEEE Trans Image Process 11:146–158
3. Tamura H, Mori S, Yamawaki T (1978) Texture features corresponding to visual perception. IEEE Trans Syst Man Cybern 6
4. Landy MS, Graham N (2004) Visual perception of texture. Vis Neurosci 1106–1118. MIT Press
5. Mallat S (1989) A Theory for Multiresolution Signal Decomposition: the Wavelet Representation. IEEE Trans Pattern Anal Mach Intell 11(7):674–693

Chapter 11
Histogram Equalization

11.1 Histogram

A histogram is a graphical representation of the distribution of numerical data. It is an estimation of the probability distribution of a continuous variable and was first introduced by Karl Pearson [1]. The histogram is a type of graphic. To build a histogram, the first step is to divide the entire range of values into a series of intervals–and then count how many values fall into each interval. The intervals in the histogram are usually specified as consecutive, non-overlapping intervals. For the most part, the images do not show a distribution.

In this first part, some examples of characteristic distributions of discrete random variables will be presented [2].

In a histogram of an image the ox–axis (horizontal axis) represents intervals of values and the oy–axis (vertical axis) represents the frequency values corresponding to each interval.

From a statistical point of view, we can consider the value of each pixel as a particular realization of a random variable associated with grey levels, in which case the histogram represents the probability density function [3].

The linear histogram of an image $U[M x N]$ represents

$$h(i) = number\ of\ pixels\ with\ grey\ level\ u\ in\ image\ U$$

Basically, histogram calculation involves going through the image pixel by pixel and counting the number of grey levels encountered (Fig. 11.1).

11.2 Histogram Equalization

In any image depending on the predominant grey levels or absent grey levels the histogram will have a different shape; most images do not show a uniform distribution of grey levels. Enhancement operations are applied to improve the visual perception

© The Author(s), under exclusive license to Springer Nature Switzerland AG 2021
M. Borda et al., *Randomness and Elements of Decision Theory Applied to Signals*,
https://doi.org/10.1007/978-3-030-90314-5_11

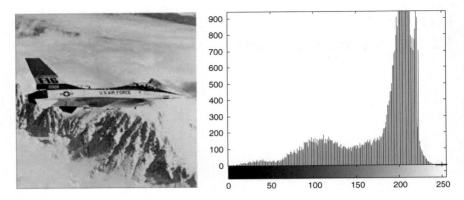

Fig. 11.1 Image and image histogram

of images and aim to redistribute the existing grey levels in the image so that they cover the full range of available values (usually between 0 and 255). Histogram equalization consists of this distribution of grey levels in a new but uniform distribution. The purpose of histogram equalization is to obtain a uniform distribution of grey levels and the resulting image will show an improvement in contrast.

Histogram equalization, from a mathematical point of view, is a problem of transforming a random distribution (described by the histogram of the original image) into a uniform distribution. Implementing histogram equalization therefore involves determining a scalar function of one variable (which changes the values of the grey levels).

The first-order distribution function for image $U[MxN]$ is calculated:

$$H_{cumulative}(i) = \sum_{j=0}^{i} h(j), i = 0, 1, \ldots, L_{Max}$$

and is called the cumulative histogram [4].

Histogram equalization is achieved by changing the value of each pixel in the image through a function that is obtained by rescaling the values of the cumulative histogram of the image.

For an image $U[MxN]$ the histogram equalization function is:

$$f_{equalization}(i) = L \sum_{j=0}^{i} h(j) = L \cdot H_{cumulative}(i), \ \forall i \in \{0, 1, \ldots, L_{Max}\}$$

Histogram equalization is an automatic method, requires no parameters to be set and is adaptive, resulting in an image with better contrast and therefore, an improved image for its visual appearance [5].

11.3 Histogram of an Image and Histogram Equalization in Matlab

In Matlab there are default functions for calculating and displaying the histogram of an image. If the image is denoted by I, then the function imhist (I) is used to display the histogram of this image.

An image can be read in Matlab using the function:I = imread (filename).

To go through the steps to do histogram equalization in Matlab there is a function implemented. If the image is denoted by *I*, and the histogram of the image is denoted by *hgram*, then to obtain histogram equalization apply J = histeq (I, hgram). In *J* we get the image with an equalized histogram and improved contrast [6].

Example 11.1 We will try to walk through the histogram equalization process and display each sample ("bin") before equalization and the value it becames after histogram equalization. We have the following Matlab code:

```
clear all;close all;clc;
GIm=imread('airplane.bmp');
numofpixels=size(GIm,1)*size(GIm,2);
figure,imshow(GIm);
title(Original image');
HIm=uint8(zeros(size(GIm,1),size(GIm,2)));
freq=zeros(256,1);
probf=zeros(256,1);
probc=zeros(256,1);
cum=zeros(256,1);
output=zeros(256,1);
% req counts the number of occurrences of each pixel
value.
% The probability of occurrence of each value is cal-
culated
 by probf.
 for i=1:size(GIm,1)
     for j=1:size(GIm,2)
         value=GIm(i,j);
         freq(value+1)=freq(value+1)+1;
         probf(value+1)=freq(value+1)/numofpixels;
     end
```

```
end
sum=0;
no_bins=255;
% The cumulative distribution is calculated.
for i=1:size(probf)
    sum=sum+freq(i);
    cum(i)=sum;
    probc(i)=cum(i)/numofpixels;
    output(i)=round(probc(i)*no_bins);
end
for i=1:size(GIm,1)
    for j=1:size(GIm,2)
            HIm(i,j)=output(GIm(i,j)+1);
    End
end
figure,imshow(HIm);
title('Histogram equalization ');
% The result is shown in the table
figure('Position',get(0,'screensize'));
dat=cell(256,6);
for i=1:256
dat(i,:)={i,freq(i),probf(i),cum(i),probc(i),output(i
)};
  end
     columnname =    {'Esantion', 'Histogram', 'Proba-
bility', 'Cumulative Histogram,'CDF','Out'};
  columnformat = {'numeric', 'numeric', 'numeric', 'nu-
meric', 'numeric','numeric'};
  columneditable =  [false false false false false
false];
  t = uitable('Units','normalized','Position',[0.1 0.1
0.4 0.9], 'Data', dat,'ColumnName', columnname,
'ColumnFormat', columnformat, 'ColumnEditable', col-
umneditable, 'RowName',[]);
     subplot(2,2,2); bar(GIm);
     title('Before histogram equalization');
     subplot(2,2,4); bar(HIm);
   title('After histogram equalization');
```

After running the code, we get the original image and the image resulting after histogram equalization (Fig. 11.2).

Example 11.2 We will calculate the histogram equalisation again and visualise the results.

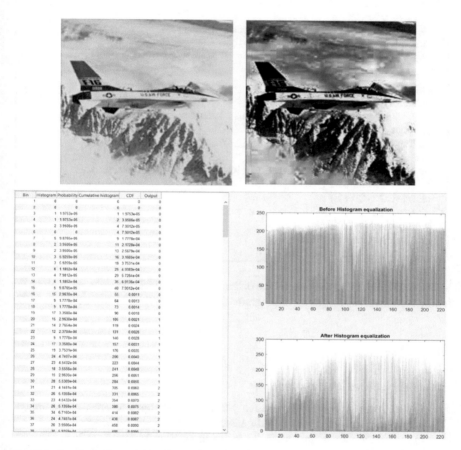

Fig. 11.2 (top-left) The original image, (top-right) the image obtained after applying histogram equalization and (bottom) the table with each sample in the histogram and the value that becomes after applying histogram equalization

```
close all;clear all;
% Read a grayscale image as a matrix mxn
A=imread('tire.tif');
figure,imshow(A);
figure;imhist(A);
% Specifies the interval in which samples are[0 255]
bin=255;
% Calculation of image histograms.
Val=reshape(A,[],1);
Val=double(Val);
I=hist(Val,0:bin);
figure;stem(I);
axis([0 255 0 2500]);
% Divide the result by the number of pixels
Output=I/numel(A);
% The cumulative amount is calculated
CSum=cumsum(Output);
% The transformation S=T(R) is done where S and R are
in the interval[ 0 1]
HIm=CSum(A+1);
% Convert image to uint8 for showing
HIm=uint8(HIm*bin);
figure,imshow(HIm);
figure; imhist(HIm);
```

After running the code, we get the original image and its histogram as well as the image resulting from applying histogram equalization and its histogram (Fig. 11.3).

11.4   Tasks

Based on the above Matlab code:

1. Create a new Matlab file in which you write code lines to open a new image, display the image and display the histogram of the opened image.
2. In the previously created Matlab file, using existing Matlab functions calculate the histogram equalization. Display the resulting image and the histogram of the new image. Comment on what happens to the image and how the histogram was transformed.

11.5   Conclusions

In this chapter we present the histogram of an image, histogram equalization applied to images and the purpose of applying histogram equalization. After the presentation of the theoretical part, you can find applications, functions and Matlab code for histogram equalization applied on images.

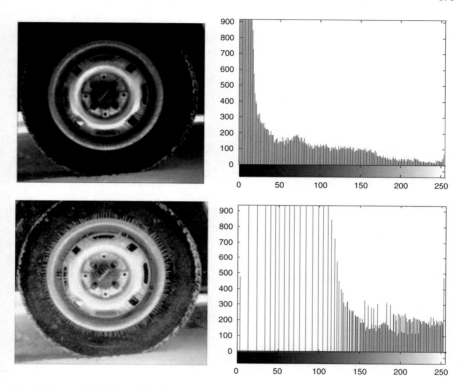

Fig. 11.3 (top-left) The original image; (top-right) original image histogram, (bottom-left) the image after applying histogram equalization; (bottom-right) modified image histogram

References

1. Pearson K (1895) Contributions to the mathematical theory of evolution. II. Skew variation in homogeneous material. Philos Trans R Soc A Math Phys Eng Sci 186:343–414
2. Howitt D, Cramer D (2008) Statistics in psychology. Prentice Hall
3. MIV Imaging Venture Image and signal processing laboratory. www.miv.ro Accessed 12 July 2021
4. Ciuc M, Vertan C (2005) Prelucrarea statistica a semnalelor. Ed. MatrixRom
5. Vlaicu A (1997) Prelucrarea imaginilor digitale. Ed. Albastra, Cluj-Napoca
6. Laerd Statistics https://statistics.laerd.com/statistical-guides/understanding-histograms.php Accessed 14 May 2021

Chapter 12
PCM and DPCM–Applications in Image Processing

12.1 Pulse Code Modulation

The concept of Pulse Code Modulation (PCM) was first introduced in 1938 by Alec Reeves, ten years before Shannon's communications theory and transistor invention, but too early to demonstrate its importance. It is the basic principle of digital communications involving a conversion of the information-carrying signal $x(t)$ from analogue to digital (ADC–analogue to digital conversion). The signal generated is a digital one, characterised by a decision rate (bit rate), so PCM is an information representation code, it is a digital representation of an analogue signal $x(t)$.

PCM generation is illustrated in the block diagram below.

At the receiver, the processing is reversed, and represents a digital to analogue conversion (DAC) followed by a low pass (LPF) of the recovered samples x_{qk}.

As shown in Fig. 12.1, PCM generation involves:

- low pass filtering the analogue signal x(t) by an anti-aliasing analogue filter whose function is to remove all frequencies above f_M (the maximum frequency of $x(t)$)
- sampling of low-pass filtered signal at a rate $f_S \geq 2f_M$ (sampling theorem); at this point, the analogue signal $x(t)$ is discretized in time, resulting in the PAM signal (Pulse Amplitude Modulation), x_k being the values of $x(t)$ at discrete moments kT_S, where $T_S = \frac{1}{f_S}$ is the sampling period.
- quantization: each sample x_k is converted into a finite number of possible values of the amplitude (q is the number of quantization levels) x_{qk}.
- encoding: the quantized samples x_{qk} are encoded using a binary alphabet ($m = 2$), each quantized sample being represented, usually, by a binary code word of length n, called a PCM word.

Thus, the infinite number of possible amplitude levels of the sampled signal is converted into a finite number of possible PCM words. If n is the length of the PCM word, the total number of possible distinct words that can be generated (possible amplitude values of the quantized sample) are $q = 2^n$.

© The Author(s), under exclusive license to Springer Nature Switzerland AG 2021
M. Borda et al., *Randomness and Elements of Decision Theory Applied to Signals*,
https://doi.org/10.1007/978-3-030-90314-5_12

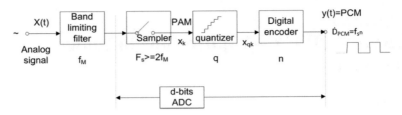

Fig. 12.1 Pulse code modulation (PCM) block scheme

The step size (quantum) Δ under the assumption that we have uniform quantization ($\Delta = constant$) is $\Delta_u = \frac{X}{q}$ for a unipolar signal $x(t) \in [0, X]$, and for a bipolar signal $\Delta_b = \frac{2X}{q}$, $x(t) \in [-X, X]$, where $|X|$ is the maximum level of the signal $x(t)$ and q denotes the quantization level.

If the number of quantization levels is q, then the PCM word length is $n = ldq \in \mathbb{Z}$, where $\mathbb{Z}$ is the integer set [1]. An example of PCM coding is presented in Fig. 12.2.

Concluding, PCM generation implies:
- band-limiting filtering;
- sampling of $x(t)$ with $f_S \geq 2f_M$;
- quantization of samples with q levels
- encoding (for uniform encoding) the q levels using words of length
  $n = ldq \in \mathbb{Z}.$

Remark

PCM is used as the standard form of digital sound in computers, compact discs, digital telephony, and other digital audio applications. In a PCM stream, the amplitude of the analogue signal is sampled regularly at uniform time intervals and each sample is quantized to the nearest value at a decision level.

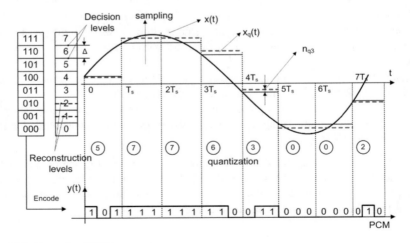

Fig. 12.2 Example of PCM generation

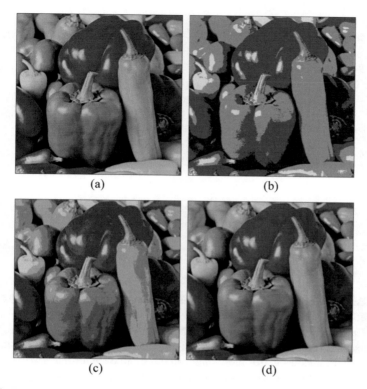

Fig. 12.3 Example of uniform quantization using PCM: **a** original image: **b** image quantized with 2 bits/pixel; **c** image quantized with 3 bits/pixel; **d** image quantized with 5 bits/pixel

In image processing, PCM coding is used to encode pixels (Fig. 12.3).

In image processing, the quantizer output is encoded with words of fixed length (typically 8 bits/word). In medical images or video signals for broadcast television, we can have 10–12 bits/pixel.

Example

12.2 Differential Pulse Code Modulation

Differential pulse code modulation (DPCM) is a procedure for converting an analogue signal into a digital signal in which samples are taken from an analogue signal and then the difference is made between the actual value of the sample and its predicted value (the predicted value is based on a previous sample or samples). This is quantized and then encoded to obtain a digital value (Fig. 12.4).

Fig. 12.4 Illustration of
DPCM principle

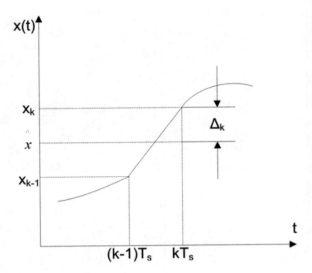

DPCM codeword represents the difference between samples, unlike PCM, where codewords represent a sample value.

The basic concept of DPCM is the encoding of a difference [2]. This process is because most source signals show a significant correlation between successive samples, so the encoding uses redundancy in the sample values that involves a lower bit rate. In PCM, the $x(t)$ signal, voice signal, video signal or others is sampled at a frequency slightly higher than the Nyquist frequency: $f_S = 2f_M$.

For example, in digital telephony with the maximum frequency $f_M = 3, 4kHz$ the sampling frequency is $f_S = 8kHz$. These samples are very correlated, the signal $x(t)$, does not change fast from one sample to the next one (it is a memory source). The corresponding PCM signal is therefore highly redundant. The basic idea for a differential PCM (DPCM) is to remove this redundancy before encoding (DPCM).

Example

In DPCM it is quantized the difference Δ_k between the sample x_k and the estimated value of it $\hat{x}_k$ obtained from prediction from the previous samples of the signal $x(t)$. The prediction is possible only if $x(t)$ has memory, its statistical characteristics being known. For a pure random signal, the prediction is impossible:

$$\Delta_k = x_k - \hat{x}_k$$

DPCM makes a lossy source compression, caused by the quantizing noise; this loss, for human users, if is under the thresholds of human audio sense (HAS) or human visual sense (HVS), is without quality loss (Fig. 12.5).

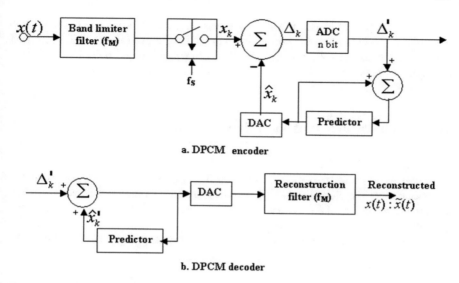

a. DPCM encoder

b. DPCM decoder

Fig. 12.5 DPCM decoder

For a linear prediction we have:

$$\hat{x}_k = \sum_{i=0}^{r} a_i x_{k-1-i}$$

r being the prediction order and a_i coefficients chosen to minimize the prediction mean square error Δ_{kmin}^2.

If $r = 0$ (zero order prediction) then $\hat{x}_k = x_{k-1}$ and $\Delta_k = x_k - x_{k-1}$ [1].

DPCM applied on signals that have successive samples correlated with each other leads to good compression factors.

Images and video signals are examples of signals that have the correlation mentioned above. In images, this correlation exists between neighbour pixels, in video signals, the correlation exists both between neighbour pixels and between the same pixels in consecutive frames.

For video signals, the DPCM compression method can be performed for intra-frame encoding (in the same frame) and inter-frame encoding (in successive frames). Intra-frame encoding exploits spatial redundancy and inter-frame encoding exploits temporal redundancy.

In intra-frame encoding, the difference Δ is calculated between pixels of the same frame, while in inter-frame encoding the difference Δ is calculated between the value of the same pixel in two consecutive frames. In both cases (intra- and inter-frame encoding) the value of the current pixel is predicted using the previously encoded neighbour pixels.

In practice, DPCM is used in lossy compression techniques, such as higher difference quantization leading to shorter codewords. This is used in the JPEG standard and in adapted DPCM (ADPCM), a method often used in audio compression. An example of predictive spatial compression scheme is presented in Fig. 12.6.

12.3   Applications and Matlab Examples

Encoding using DPCM [3]:
>   indx = dpcmenco(sig, codebook, partition, predictor)
>   dpcmenco – implements DPCM encoding to encode a vector denoted with sig.
>   Decoding using DPCM [3]: sig = dpcmdeco(indx, codebook, predictor)
>   dpcmdeco – implements DPCM decoding to decode a vector denoted by indx

12.3.1   Study of a PCM System

Example 12.1 Let a signal be continuous in time and amplitude. We encode this signal so that it can be transmitted on a channel. We use PCM encoding.

```
% PCM
clc;close all;clear all;
n=input('Enter the number of bits for a PCM word., n :  ');
n1=input('Enter the number of samples for a time period : ');
L=2^n;
% Sampling operation
x=0:2*pi/n1:4*pi;
% n1 number of samples selected
s=8*sin(x);
subplot(3,1,1);
plot(s);
```

Fig. 12.6 Example of a predictive spatial compression scheme: **a** the original image; **b** the prediction error image for a 6 bit/pixel representation using only the previous neighbour; **c** the decoded image corresponding to the error image in **b**; **d** the prediction error image for a 6 bit/pixel representation using the previous neighbour and 2 neighbours on the previously decoded top line; **e** the decoded image corresponding to the error image in **d**; **f** the prediction error image for a 1 bit/pixel representation using only the previous neighbour; **g** the decoded image corresponding to the error image in **f**; **h** the prediction error image for a 6 bit/pixel representation using the previous neighbour and 2 neighbours on the previous decoded top line; **i** the decoded image corresponding to the error image in **h**

Fig. 12.6 (continued)

```
title('Analogic Signal');
ylabel('Amplitude--->');
xlabel('Time--->');
subplot(3,1,2);
stem(s);grid on;  title('Sampled signal');  ylabel('Amplitude--->');
xlabel('Time--->');
%  Quantification process
 vmax=8;
 vmin=-vmax;
 del=(vmax-vmin)/L;
 part=vmin:del:vmax;
% level between vmin and vmax with the difference of del
 code=vmin-(del/2):del:vmax+(del/2);
% contains quantized values
 [ind,q]=quantiz(s,part,code);
% quantization process
% ind contains the value index and q contains the quantized value
```

```
l1=length(ind);
 l2=length(q);
for i=1:l1
if(ind(i)~=0)     % to bring the index between 0 and N
        ind(i)=ind(i)-1;
    end
    i=i+1;
 end
  for i=1:l2
     if(q(i)==vmin-(del/2)) % to bring quantified values between
decision levels
          q(i)=vmin+(del/2);
     end
 end
 subplot(3,1,3);
 stem(q);grid on;                % Display quantified values
 title('Quantized Signal');
 ylabel('Amplitude--->');
 xlabel('Time--->');
%  Encoding process
 figure
 code=de2bi(ind,'left-msb');     % conversion from decimal to binary
 k=1;
for i=1:l1
    for j=1:n
        coded(k)=code(i,j);% conversion of the encoding matrix into a
line vector
        j=j+1;
        k=k+1;
    end
    i=i+1;
end
 subplot(2,1,1); grid on;
 stairs(coded);       % display encoded signal
axis([0 100 -2 3]);  title('Encoded Signal');
 ylabel('Amplitude--->');
 xlabel('Time--->');
%   Demodulation of a PCM signal
  qunt=reshape(coded,n,length(coded)/n);
index=bi2de(qunt','left-msb');% index in decimal form
 q=del*index+vmin+(del/2);        % quantized values
 subplot(2,1,2); grid on;
 plot(q);                 % Display demodulated signal
```

The results are presented in Fig. 12.7:

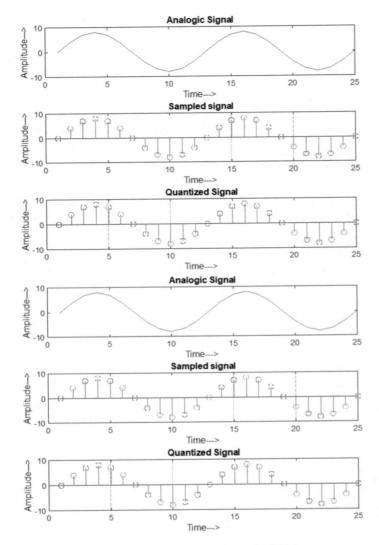

Fig. 12.7 Illustrating the analogue signal encoding process using PCM

12.3.2 Study of a DPCM system

Example 12.2 Let an image be grayscale as in the example below. We want to encode the image using the DPCM algorithm. We will do the decoding and we will display the result.

```
close all; clear all; clc;
predictor = [0 1]; % y(k)=x(k-1)
partition = [-1:.1:.9];
codebook = [-1:.1:1];
y=imread('Lena256.bmp');
m=size(y,1); n=size(y,2);
figure; imshow(y);
y=double(y);
x = y(:);
t = linspace(1, 2*pi, length(x));
% quantization x using DPCM.
encodedx = dpcmenco(x,codebook,partition,predictor);
% Attempt to recover x from the modulated signal
decodedx = dpcmdeco(encodedx,codebook,predictor);
figure;  plot(t,x,t,decodedx,'--')
dx =decodedx(:);
[psnrtest]=psnr(x,dx)
  legend('Original signal','Decoded
signal','Location','NorthOutside');
A=reshape(dx,m,n);
figure; imshow(uint8(A));
```

The obtained results for the image *Lena* are presented in Fig. 12.8.

12.4 Conclusions

The chapter presents pulse code modulation (PCM) and differential pulse code modulation (differential PCM or DPCM) with applications in image processing. For the application part, an illustration on images of the PCM encoding process and a DPCM encoding process is shown.

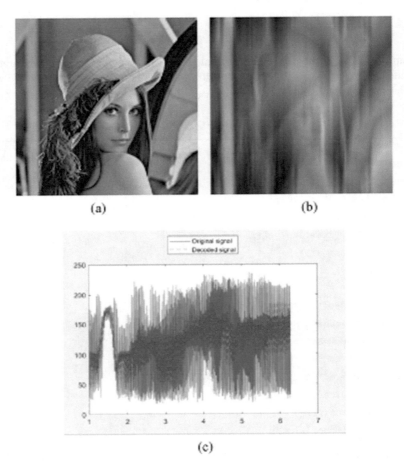

Fig. 12.8 Results. **a** the original image; **b** the image obtained after decoding; **c** the representation of the original signal and the decoded signal

References

1. Borda M (2011) Fundamentals in information theory and coding. Springer
2. Rechnernetze http://einstein.informatik.uni-oldenburg.de/rechnernetze/dpcm.htm Accessed 29 May 2021
3. Matlab https://www.mathworks.com/ Accessed 15 Sept 2021

Chapter 13
Nearest Neighbour (NN) and k-Nearest Neighbour (kNN) Supervised Classification Algorithms

13.1 Unsupervised and Supervised Classification of Data

Data classification approaches aim to group data into homogeneous classes according to a chosen dissimilarity or homogeneity criterion (distance, statistic inclusion in a specific probability density function etc.) and the representation of the grouped data in a labelled format. The grouping operation can be performed in an unsupervised or supervised manner. In non-supervised classification, data is presented to the classification system under the form of a set of vectors containing representative characteristics, but without knowing a priori their belonging to a predefined set of classes. The classification system aims in this case to partition the set into homogeneous classes. Supervised classification involves human intervention and divides the classification process into two stages: training and prediction of a categorial label [1]. In the training stage, multiple annotated data samples are presented to the system, labelled, and grouped in M classes. This stage aims at learning a classification rule or a classification model that is capable to predict the categorical label associated to one of the M classes whenever an unknown data sample is presented at the input. Supervised classification systems currently have applications in extremely varied fields: data mining techniques, pattern recognition, signal and image processing, remote sensing, indexing and automatic data search, artificial intelligence.

13.2 NN and kNN Classification Algorithms

NN and kNN classification algorithms are simple and intuitive algorithms that can easily be parameterized for a certain number of classes. The training stage involves a simple storage of the labelled training data. Let:

$$x = (x_1, x_2, \cdots, x_p)^T \subset R^p$$

be a p-dimensional feature vector and:

© The Author(s), under exclusive license to Springer Nature Switzerland AG 2021
M. Borda et al., *Randomness and Elements of Decision Theory Applied to Signals*,
https://doi.org/10.1007/978-3-030-90314-5_13

$$c(x) = j, \ j \in \{1, 2, \cdots, L\}$$

the annotated class associated to observation x.

The set $Y = \{(\boldsymbol{x}_i, c(\boldsymbol{x}_i)), i = 1, 2 \cdots N\}$ represents the set of training vectors.

For each new observation $y = (y_1, y_2, \cdots, y_p)^T$ the NN or kNN classification algorithms compute a distance function $d_j(\boldsymbol{y}) = d(\boldsymbol{x}_j, \boldsymbol{y})$ between this new data sample and all the vectors from the training set. The used distance functions are usually of Minkovski type:

$$d(x_j, y) = \left(\sum_{j=1}^{n} |x_i - y_i|^n \right)^{1/n} , n = 1, 2$$

In the categorial prediction stage, the NN and kNN algorithms determine the closest k vectors from the training set. The principle is illustrated in Fig. 13.1 for a binary classification problem based on the Euclidian distance similarity criterion.

Remark

- The NN algorithm is a particular case of the kNN algorithm ($k = 1$)
- k is a parameter of kNN classifiers that depends on the application. In practice, the optimal value of k is determined by trial and error like approaches or using *cross-validation* techniques in the training stage
- In the training stage, known data samples are further divided into a *training dataset* and a *test dataset*. The model is built using the training dataset and evaluated on the test dataset using cross-validation methods. The evaluation metric is usually the accuracy defined by the number of correct predictions and the total number of samples in the test dataset
- One of the most used cross validation methods is the one in which K randomly split equal part disjoint sets (called folds) of the training set are used. K-1 subsets are used for the training stage and the remaining one is used in the testing/prediction stage (K–*fold validation*). The procedure is repeated for each fold as a test fold and the maximum mean accuracy indicates the optimal value for k
- An additional data normalization step can be used for both the training and prediction stages. One of the most used methods is the Z-score normalization approach [2] that aims to rescale the space of parameters so that each characteristic in (1) can be considered as being extracted from a probability density function whit a zero mean and unitary standard deviation. Each component of the N training vectors is normalized in this case according to the relationships:

$\mu_i = \frac{1}{N} \sum_{i=1}^{N} x_i$

$\sigma_i = \sqrt{\frac{1}{N-1} \sum_{i=1}^{N} (x_i - \mu_i)^2}$

$x_i^{norm} = \frac{x_i - \mu_i}{\sigma_i}$

- The last relationship also defines the normalization rule for the feature vector associated with a new observation.

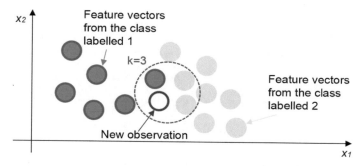

Fig. 13.1 Prediction of a categorical label using the kNN algorithm (k = 3)

13.3 Support for NN and kNN Algorithms in Matlab

The Matlab programming language provides facilities for the implementation of NN or kNN classification algorithms. The Matlab's dedicated function for the training stage is called fitcknn(.) [Matlab].

The function takes as main parameters two tables X (the set of training vectors) and Y (the set of labels associated to the training vectors).

Their use will be illustrated based on an example.

Example

Let a training set consist in 10 two-dimensional vectors $(1,-2)^T$, $(1.25,-1.6)^T$, $(1.5,-1.75)^T$, $(1.75,-1.5)^T$, $(2,-1.75)^T$, $(2.25,-1)^T$, $(4,0)^T$, $(4.25,-0.25)^T$, $(4.5,0.1)^T$, $(4.75,0.25)^T$.

Such a data structure can be defined using the following Matlab variable declaration:

```
X = [1,1.25, 1.5, 1.75, 2, 2.25 4, 4.25, 4.5,
4.75; -2,-1.6 -1.75, -1.5, -1.75,-1, 0, -0.25,
0.1, 0.25];
```

Graphical visualization of the distribution of the data contained in the training set can be performed using Matlab's graphic display facilities (the scatter (x,y) function) passing the training vectors as parameters of the function (scatter(X(1,:), X(2,:)). The result is illustrated in Fig. 13.2.

Labeling the training set involves manually assigning labels to each of the 10 vectors. In the case of the example considered this can be done by using a proximity relationship between the vectors represented in the space of the characteristics. The Matlab instruction that can be used in this regard is a simple instruction that defines a one-dimensional table with initialization:

```
Y = [1, 1, 1, 1,1,1, 2, 2, 2, 2];
```

Creating the kNN model for a classification operation based on the 3 closest neighbors in Matlab involves the use of the following sequence of operations:

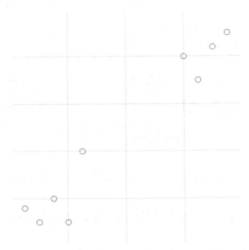

Fig. 13.2 Graphical visualization of vectors in the training set

```
training=transpose (X);
k=3;
mdl = fitcknn(training,Y,'NumNeighbors',k);
```

The mdl variable created by Matlab contains all the properties of the kNN model to be used (Fig. 13.3).

The distance metric was not specified in the creation of the model. In this case, the default distance function that will be used is the Euclidean one. The created model is a general one: it allows the assignment of weights to individual observations,

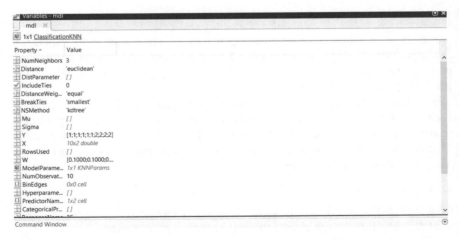

Fig. 13.3 Visualisation of the created data structure for kNN classification in Matlab

```
>> label=predict(mdl,y)

label =

    1

>>
```

Fig. 13.4 Result of the categorical prediction for the observation y = (2.5,−1)

normalization of data, definition of a decision rule in case of obtaining an equal number of representatives from the annotated classes, rules for cross-validation, the definition of new metrics to determine the degree of similarity, etc. [3].

The classification of a new observation in the prediction stage involves the use of a simple instruction of the following type:

```
label = predict (mdl, y)
```

with the mdl data structure representing the previously created model and with y representing the new observation that is intended to be classified. The result of using these operations for a new observation y = (2.5,−1) is illustrated in the following Fig. 13.4.

13.4   Tasks

1. Write a Matlab app that integrates the previous example. Which is the result of the categorial prediction stage for the observation $y = (3,-0.5)^{\mathrm{T}}$ if $k = 1$? What if $k = 3$? Explain the eventual differences.
2. Generalize the example in the previous subsection for a classification problem with 3 separable classes having 5 feature vectors per class in the training set. For the prediction stage choose new observations close to, respectively, the feature vectors whose components are given by the average of the components in each class—the μ_i values in (4). (*Indication*: if X is the set of training vectors, the Matlab code $U = mean(X)$ returns the mean values of the components in a vector U. Normalization is carried out on the training matrix having the training vectors as lines). Write down the obtained results for different values of k.
3. Use the *Fisher's iris* dataset included in the existing Matlab installation on the computers in the lab and the explanations on the NN and kNN classification algorithms to study their classification performances on this dataset. The dataset includes 150 feature vectors for images of iris [3], grouped in 3 annotated classes. Each vector represents an iris image for which 4 descriptors are computed. Build the classification model for different values of k, for the distance functions and using data normalization techniques. Write down in a table the classification performance on the training dataset using *the K-fold* validation method. (*Indication*: the validation step allows to determine a loss (cost)

function by minimizing the error between the built model and the predicted response. The dedicated Matlab function that can be used for this purpose is *loss (validated_ KNN_model,…)* [3]).

13.5   Conclusions

The theoretical fundamentals associated with simple and intuitive supervised classifiers (NN and kNN) are presented. The theoretical aspects are accompanied by practical examples of the use of classifiers in Matlab, with the optimization of the parameters.

References

1. Smola A, Vishwanathan SVN (2008) Introduction to Machine Learning. Cambridge University Press
2. Abdi H (2010) Normalizing Data. In: Salkind Neil (ed) Encyclopedia of Research Design. Sage, Thousand Oaks, CA
3. Matlab www.mathworks.com Accessed 14 Sept 2021

Chapter 14
Texture Feature Extraction and Classification Using the Local Binary Patterns Operator

14.1 Automatic Image Classification System

Texture classification is an important field in the image processing domain which involves the allocation of a predefined label to an input image. In this work, we will present the basic principles of an automated system that includes algorithms for classifying images containing textures. This type of system can be useful in industry, precision agriculture, in the biomedical domain, and many others.

Even if image classification can be used in a variety of applications and areas, this is not an easy task because images are not always uniform and can exhibit changes in scale, illumination, rotation, and are also affected by different types of noise. However, even if there is a multitude of application areas, two steps are always considered in the process of classifying images and textures in a supervised machine-learning manner: the training step and the classification one (the test phase). Figure 14.1 shows a block diagram that presents in detail how these steps are correlated and their functionality.

For the training phase, a training dataset is used. In this set, for each sample, there is also given the corresponding category. In this step, the training textures are analyzed by using an algorithm for feature generation. The result is a train feature vector. A quantitative and compact description of the images is obtained by generating the most relevant texture information. Having as input the categories and the extracted features from the train feature vector, the machine learning algorithm determines a model.

In the classification step, the input of the system is a new image, unseen in the training process. The textural content of the unknown sample is extracted by considering the same texture analysis technique. The classification algorithm is then used to compare the features of the test image with those obtained for the training dataset. The new sample is then assigned to the nearest category according to a certain rule, depending on the chosen classification technique.

M. Borda et al., *Randomness and Elements of Decision Theory Applied to Signals*, https://doi.org/10.1007/978-3-030-90314-5_14

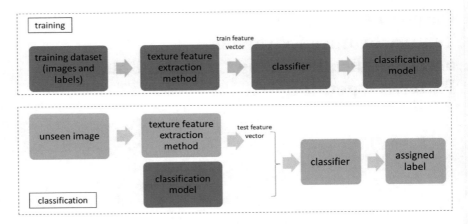

Fig. 14.1 Schematic diagram of a supervised texture classification system

14.2   Feature Extraction

14.2.1   Generalities

The feature extraction process implies extracting features from the analyzed images. Features are attributes that are discriminative, being essential in a classification system as its performance is highly impacted by the description of the images generated through the feature extraction step. Therefore, the generated attributes should contain the most relevant information such as to be able to distinguish between different categories. This involves obtaining a low variability within the same class and high variability between different image categories.

The number of extracted features should also be reduced in order to limit the feature vector size. Additionally, instead of having lots of data with redundant information, by considering fewer but meaningful features, a compact image characterization is obtained. This is also more efficient in terms of storage and processing time. Moreover, the extracted features must provide invariance capabilities to different transformations of the analyzed images (for example rotation, observation scale, and illumination changes).

The feature extraction step involves generating features for each considered sample. The features can be scalar numbers or histograms and they are stored in a feature vector. Many descriptors, global or local, can be used to extract textural features. This work presents in the following one of the most popular and efficient texture feature extractors, the Local Binary Patterns.

14.2.2 The Initial Local Binary Patterns Operator

The Local Binary Patterns (LBP) operator is a texture analysis descriptor proposed by [1]. The most important advantages of this feature extraction method are its illumination invariance, short computation time, and its efficiency in the pattern description field, being used in various applications with great success: face analysis and recognition, texture classification, and shape localization.

LBP operates on grayscale images. For calculating the LBP code for a given pixel, some of its neighbors are considered. They form a small neighborhood around the analyzed central pixel, x_c. This neighborhood contains P pixels. For the original LBP operator proposed in [1], $P = 8$ neighbors are used.

If the analyzed images are RGB, they must be converted to grayscale. Then, for each pixel in the grayscale image, we need to select a neighborhood of size 3×3 surrounding the central pixel. Figure 14.2 shows an example that illustrates the thresholding operation used in LBP for a given neighborhood where the intensity of the central pixel is 40.

The value of each pixel of the input image is compared to the intensities corresponding to its 8 neighboring pixels: if the intensity of the neighboring pixel is greater than or equal to the intensity value of the central pixel, the value 1 is noted and 0 otherwise as expressed in:

$$x'_k = \begin{cases} 1, & \text{if } x_k \geq x_c \\ 0, & \text{if } x_k < x_c \end{cases}$$

where $x_k, k = 1 : 8$ denotes the values of the neighboring pixels' intensities before thresholding and $x_k', k = 1 : 8$ are the values corresponding to the neighboring pixels after the thresholding process.

The concatenation of all of these binary values yields an 8-bit binary number for each pixel location of the input textured image. This binary value is then transformed to a decimal number denoted by n by considering:

$$n = \sum_{k=1}^{8} 2^{k-1} * x'_k.$$

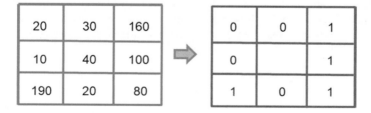

Fig. 14.2 Thresholding principle in LBP

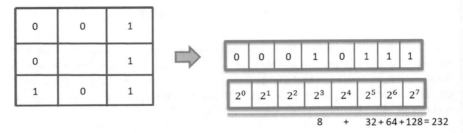

Fig. 14.3 Conversion to decimal

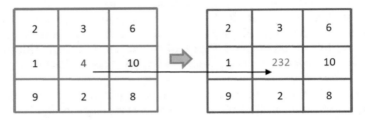

Fig. 14.4 Storing the decimal number in the output matrix

Figure 14.3 shows the conversion to decimal for the same example.

The obtained decimal number is 232. Figure 14.4 shows how this decimal number is stored in the output matrix.

The expression used to compute the original LBP operator is:

$$LBP_8 = \sum_{k=1}^{8} s(x_k - x_c) * 2^{k-1},$$

where $s(x)$ is given by: $s(x) = \begin{cases} 0, & \text{if } x < 0 \\ 1, & \text{if } x \geq 0 \end{cases}$ and the subscript 8 indicates the considered number of neighbors.

This process of thresholding, aggregation of bit arrays, and storage of the output decimal value in the LBP code matrix is then considered for all pixels in the input image. The last step is to calculate a histogram of the values contained in the output LBP matrix. Since there are 8 neighbors, there are $2^8 = 256$ possible patterns, so the histogram has 256 bins. This histogram is considered as feature vector.

14.2.3 The Improved Version of the Local Binary Patterns

The LBP operator was improved by the work of [2]. The new variant uses neighborhoods of any size, not just 3×3. This improvement was introduced so that the LBP operator can generate texture features acquired at various scales. In this case, the neighborhood is composed of a set of sample points that are uniformly placed on a circle. Since any circle radius and number of sample points/neighbors can be used, there are situations in which certain values do not exactly correspond to full pixel coordinates. In this case, their values are estimated by bilinear interpolation. In what follows, the notation (P, R) will be used to denote neighborhoods with P sampling points (neighbors) on a circle of radius R. Figure 14.5 shows an example of circular neighborhood.

For obtaining rotation invariance, the rotation invariant LBP operator was proposed in [2]:

$$
\text{LBP}_{R,P}^{riu2} = \begin{cases} \sum_{k=1}^{P} s\left(x_c - x_{R,P,k}\right), & \text{if } U\left(\text{LBP}_{R,P}\right) \leq 2 \\ P + 1, & \text{otherwise} \end{cases}
$$

where U is called the uniformity function and is used to measure how many bitwise transitions are encountered in a pattern, $riu2$ stands for uniform rotation invariant patterns for which $U \leq 2$ and $x_{R,P,k}$ are the values of the pixels located in a neighborhood of radius R which contains P neighboring pixels.

Fig. 14.5 Example of circular neighborhood

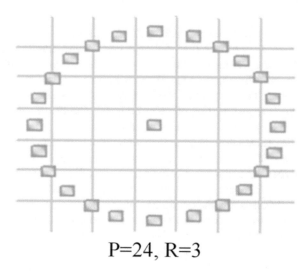

$$P=24, \ R=3$$

The number of uniform patterns in a circular neighborhood containing P neighbors is $P + 1$. An individual label is associated to each uniform pattern by considering the number of bits of 1 in that pattern. All non-uniform patterns are collected in the same class having the label $P + 1$. The feature vector is represented by the histogram composed of $P + 2$ values.

14.3   Image Classification

14.3.1   Introduction

In the supervised classification, the decision is based on the training set which consists of a set of images for which the true labels are known a priori. A machine learning technique is an algorithm that can apply to new data what it has learned from previous data. This step is called the training phase. The training dataset is utilized as input for the machine learning classifier which creates a model that is used to predict labels for future data. The second phase, known as testing, is used to assess the quality of the classification task. Therefore, as input for the classification system, a new set of images never seen before by the classifier is used. Then, the true labels are compared to those predicted by the classifier and the accuracy of the method can be computed.

14.3.2   The Support Vector Machine Classifier

The Support Vector Machine (SVM) algorithm is a well-known supervised machine learning approach that seeks the best hyperplane that segregates the image classes. Let us consider that there are n features in each sample feature vector. Therefore, in an n-dimensional space, each sample may be represented as a point. The ideal hyperplane is the one for which the image classes are separated in the n-dimensional space representation of the training samples by an interval as wide as possible. In the test step, a new sample is represented in the same space. The decision concerning the classification depends on where the new sample point is located with respect to the side of the interval. The margin is defined as the distance between the separating hyperplane and the closest data points (support vectors). The optimal hyperplane that can separate the classes is the one with the largest margin. This is the simplest SVM classifier, the maximum margin linear classifier. Let us consider a classification problem with two classes, blue squares and red circles, where each sample has two features (the hyperplane is thus a line). Figure 14.6 illustrates the maximum margin linear SVM applied to this example.

However, in practice, real data cannot be perfectly separated with a hyperplane. We want to find the plane with the largest margin but also the one that separates the

Fig. 14.6 Example of maximum margin linear SVM binary classification

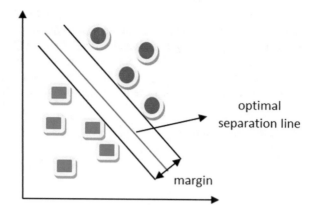

optimal
separation line

margin

classes correctly. Sometimes both conditions cannot be reached simultaneously. So, the constraints need to be relaxed, which leads to the SVM soft margin classifier. For this method, some training sample points are authorized to overstep the separation hyperplane, depending on the value of a parameter denoted by C. A low C value implies considering a margin as wide as possible even if more samples are erroneously classified. A high C value involves the opposite: the margin is allowed to be smaller if this implies a smaller number or zero misclassifications.

In addition, there are situations where data is not linearly separable. In order to solve this issue, the input space is converted into a space with a larger number of dimensions where data can be separated linearly. This is called the kernel trick. Thus, the non-separable problem is transformed into a separable one. Different kernel types can be considered: polynomial, linear, or Radial Basis Function (RBF) [3].

Another parameter that has to be chosen when using the SVM algorithm is γ. It controls how a training sample can influence the classification in terms of distance. If it is larger, it means that the measure of similarity is not so strict and the probability that distant data are similar is higher. Smaller values indicate that only very near data can be alike.

14.4   Work Description

We are going to perform the classification of images from a texture dataset, Outex_TC_00012 ("[4]," Outex Texture Database). It contains 4800 textured images divided into 24 classes. The images are rotated at various angles and are captured under two different lighting conditions [4].

We are going to use Matlab to implement the classification system described in the theoretical part. Let us consider that the dataset is stored at the location given by the variable path1 (example: 'E:/image classification/database/'). Images from a given category are stored in a separate folder whose name denotes the membership class of the images contained in that folder. The folder structure is presented in Fig. 14.7. For performing the SVM classification, the LIBSVM library is used [5]. The variable `path2` contains the path to the folder that includes the LIBSVM functions.

First of all, we are going to initialize some parameters and load the images from the database into the workspace using the following code:

Fig. 14.7 Database folder structure

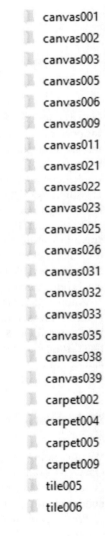

canvas001

canvas002

canvas003

canvas005

canvas006

canvas009

canvas011

canvas021

canvas022

canvas023

canvas025

canvas026

canvas031

canvas032

canvas033

canvas035

canvas038

canvas039

carpet002

carpet004

carpet005

carpet009

tile005

tile006

```
clear all
close all
path1=....;
path2=....;
addpath(path2);
LBP_features_train=[];
LBP_features_test=[];

%parameters
percentage=0.75;
SVM_param='-t 2 -c 100 -g 0.01';

% load images into image datastore
imds = imageDatastore(path1,
'IncludeSubfolders',true,'FileExtensions',
'.ras','LabelSource','foldernames');
```

Please complete the above code with the paths corresponding to the two variables: path1 and path2. The LBP_features_train and LBP_features_test will store the feature vectors corresponding to the train and test datasets, respectively. They are initialized as empty arrays. The percentage variable represents the considered percentage of training images from the entire dataset. The SVM parameters are stored in the SVM_param variable. The possible options for these parameters can be found by running the following command in the command window:

```
svmtrain
```

In our case, the considered SVM parameters are: Radial Basis Function kernel ($-t\,2$), $C = 100$ and $\gamma = 0.01$. The variable imds stores an image datastore object that is used to easily manage and collect image files. We can count the number of images per class, the total number of images, and classes using the following code extract:

```
% count the number of images in each class
labelCount = countEachLabel(imds);
% total number of images in the dataset
nr_images_total=size(imds.Labels,1);
% the number of classes in the dataset
nr_classes=size(labelCount,1);
```

The next step is to split the considered dataset into two subsets: one used for training and one used for testing. We can do this by considering the following Matlab commands:

```
% split data randomly into train and test
[imdsTrain,imdsTest] = splitEachLabel(imds,
percentage,'randomized');
% convert labels from categorical to double format
labels_train=double(imdsTrain.Labels);
labels_test=double(imdsTest.Labels);
```

By considering the option `'randomized'` randomized', the function *splitEachLabel* is used to split randomly the images from the initial database into two subsets by considering the percentage variable that corresponds to the training samples. The two subsets are: imdsTrain and imdsTest. The labels_train and labels_test variables store the corresponding correct categories in column arrays. By converting them to a double type format, category names are converted to decimal numbers from 1 to 24. They are more easily managed in this way.

The next step is the feature extraction process. We are going to loop through all training images and for each image, we are going to compute the LBP features. Matlab offers the extractLBPFeatures function that generates the L2 normalized uniform LBP histogram for a given input image:

```
% loop through all train images to extract
LBP features
for i=1:size(imdsTrain.Labels,1)
    % read the image
    img=imread(imdsTrain.Files{i});
    % verify if the image is RGB and if it is,
```

```
convert it to grayscale
    if(size(img,3)==3)
        img=double(rgb2gray(img));
    end
    % extract riu2 LBP features
    features = extractLBPFeatures(img,
'NumNeighbors',8,'Radius',1,'Upright',false);
    % store the features in a matrix
LBP_features_train=[LBP_features_train;double
(features)];
end
```

The LBP operator only works on grayscale images. Therefore, if the input images are RGB, they are converted to grayscale. For the extractLBPFeatures function, we can set the following parameters: the number of neighbors used (in the presented example, it was set to 8), the radius (set to 1 in the given example), and `'Upright'`. The `'Upright'` option is the rotation invariant flag. If it

Fig. 14.8 Example of matrix that stores the train feature vectors

		Feature 1	Feature 2	....	Feature 10
TRAIN FEATURE VECTOR MATRIX					
	Image 1	12	4656	...	67
	Image 2	54	7000	...	643
	......	....	...	...	...
	Image 3600	67	6980	...	77

is set to `false`, the function computes rotation invariant features (*riu2*). In the `LBP_features_train` variable, all train feature vectors are stored as rows. Each row corresponds to a given training image and each column corresponds to a feature. An example of such matrix is presented in Fig. 14.8. For the given Matlab code, the number of features is $P + 2 = 8 + 2 = 10$ since we consider $P = 8$ neighbors and since the *riu2* coding scheme is applied (`'Upright'` is set to `false`).

We will perform exactly the same steps for the test subset:

```
for j=1:size(imdsTest.Labels,1)
    % read the image
    img=imread(imdsTest.Files{j});
    % verify if the image is RGB and if it is,
convert it to grayscale
    if(size(img,3)==3)
        img=rgb2gray(img);
    end
    % extract riu2 LBP features
    features = extractLBPFeatures(img,
'NumNeighbors',8,'Radius',1,'Upright',false);
    % store the features in a matrix
LBP_features_test=[LBP_features_test;double
(features)];
end
```

Before the classification part, the features should be standardized because the SVM classifier is sensitive to the scale of the computed features. This means we are going to perform a Z-score normalization for sample data with mean μ and standard deviation σ:

$$Z = \frac{X - \mu}{\sigma}.$$

After the normalization, the data will have a zero mean and unitary standard deviation. The code used to perform the Z-score normalization is given below:

```
% Z-score normalization along the columns
[LBP_features_train,~,~]
zscore(LBP_features_train,0,1);
[LBP_features_test,~,~] =
zscore(LBP_features_test,0,1);
```

The next step is to perform the training of the SVM classifier in order to obtain the training model. We are going to use the svmtrain function provided by LIBSVM:

```
% build the SVM model based on the training set
model=svmtrain(labels_train,
LBP_features_train,SVM_param);
```

The svmtrain function takes as parameters the train labels, the matrix which stores the train image features (see Fig. 14.8), and the SVM parameters defined at the beginning.

The next step is the actual classification when the SVM classifier predicts the categories for the images contained in the test set. This is done by using the svmpredict function which takes as inputs the test labels, the matrix which stores the test image features, and the obtained SVM model.

```
% make the prediction for the test set
[labels_predicted,~,~]=svmpredict(labels_test
,LBP_features_test,model);
```

After performing the prediction, we are interested to compute the classification accuracy of the system:

```
% compute the accuracy
accuracy = sum(labels_predicted == labels_test)/
numel(labels_test);
```

The variable accuracy stores the obtained accuracy which is 93.33% for this particular case:

 accuracy   0.9333

Please note that the accuracy varies depending on the particular random choice of the training and test dataset.

Another important indicator that is used to describe the performance of a classification system is the confusion matrix. The rows of the matrix correspond to the real class labels whereas the columns correspond to the predicted ones. The elements on

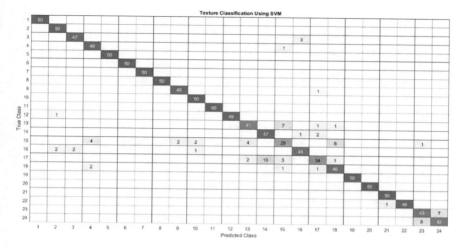

Fig. 14.9 The confusion matrix computed for the particular partition of the training and test sets

the principal diagonal represent the number of images for which the prediction is correct (the higher, the better) and the other elements indicate misclassified images. Figure 14.9 shows the computed confusion matrix for this random partition of the train and test sets. The code used to represent it is also given below. For the test dataset, by considering 75% of the images for training, there are 1200 samples for testing, so 50 samples per class. We can see from Fig. 14.9 the fact that there are classes such as category 1 for which the prediction is correct for all images (50 samples on the principal diagonal). On the other hand, for class 15, only 29 samples out of 50 are correctly classified.

```
% compute the confusion matrix and display it
[A, ~, ~]=confusion(labels_test,labels_predicted,
nr_classes);
cm = confusionchart(A);
cm.Title = 'Texture Classification Using SVM';
```

14.5 Tasks

1. Run the code from the *Work Description* section. Adjust the values of the SVM parameters and build a table containing the obtained classification accuracy values for different C and γ values. Does the classification accuracy change? Why?

   Note: typical values for C and γ range from 10^{-3} to 10^{4} (with a step of 10).

2. Write a Matlab code that uses a grid search to determine the optimal values for C and γ. You can use this range: 10^{-3} to 10^4.
3. Consider that 16 neighbors are used in the LBP feature extraction process. Run the code for this situation. How many features does the feature vector have? Does the classification performance improve? What about changing the radius to $R = 2$?
4. Change the given code such as to use the multiresolution strategy for LBP. To obtain a more discriminative operator, it would be useful to capture more texture information at different scales. The multiresolution approach proposed in [2] involves the idea of combining several $LBP_{R,P}^{riu2}$ of different (P, R) pairs by concatenation. Consider this approach and comment on the obtained results (classification accuracy and confusion matrix).
5. Perform the classification on another texture database such as Outex_TC_00013 ("[4]," Outex Texture Database). Compare the obtained results with the ones achieved previously for Outex_TC_00012.

14.6   Conclusions

This work presents the theoretical fundamentals associated to the supervised classification of textured images using the Local Binary Patterns operator and the Support Vector Machine classifier. The theoretical aspects are accompanied by practical examples developed in Matlab regarding the classification of images belonging to a popular texture dataset.

References

1. Ojala T, Pietikäinen M, Harwood D (1996) A comparative study of texture measures with classification based on featured distributions. Pattern Recogn 29:51–59
2. Ojala T, Pietikainen M, Maenpaa T (2002) Multiresolution gray-scale and rotation invariant texture classification with local binary patterns. IEEE Trans Pattern Anal Mach Intell 24:971–987
3. Burges CJC (1998) A tutorial on support vector machines for pattern recognition. Data Min Knowl Discov 2:121–167
4. Outex Texture Database. http://www.outex.oulu.fi/index.php?page=classification. Accessed 1 Nov 2018
5. Chih-Chung C, Chih-Jen L (2021) LIBSVM—a library for support vector machines https://www.csie.ntu.edu.tw/~cjlin/libsvm/#download.Accessed 5 May 2021

Chapter 15
Supervised Deep Learning Classification Algorithms

15.1 Convolutional Neural Network (CNN)

15.1.1 Introduction

First let's discuss about different applications that we use every day, but do not realize that they are implemented by using different convolutional neural networks elements. For example, different facial recognition applications on social media, or object detection in building self-driving cars, or disease detection using visual imagery in healthcare. All those applications are possible due to the convolutional neural networks (CNN).

A convolutional neural network represents a feed-forward neural network that is generally used to analyze visual images by prdocessing some data, and it's also known as a ConvNet architecture. A CNN network is used to detect and classify objects in an image. Below is a neural network that identifies two classes of animals: cats and dogs (Fig. 15.1).

A Convolutional Neural Network is a representation of deep learning algorithm which takes at the input an image, based on which one can assign importance (learnable weights and biases) to various elements in the image, in order to differentiate one from the other [1]. The pre-processing required in a ConvNet is much lower as compared to other classification algorithms. One important mention for the ConvNets is that they could learn different characteristics in the image through the different variations of filters, such as convolutional layers, max-pooling layers.

15.1.2 Layers in a CNN Network

A Convolutional Neural Network is capable to capture the spatial and temporal dependencies in an image through a number of different types of filters. The architecture performs a better fitting to the image dataset due to the reduction in the number

© The Author(s), under exclusive license to Springer Nature Switzerland AG 2021
M. Borda et al., *Randomness and Elements of Decision Theory Applied to Signals*,
https://doi.org/10.1007/978-3-030-90314-5_15

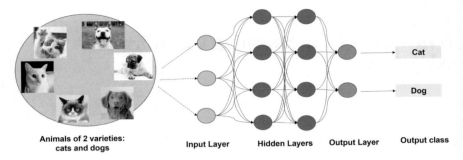

Fig. 15.1 An example of a neural network used to identify animals belonging to two classes, cats and dogs

of parameters involved for each layer in the neural network, and the reusability of weights. This means that the network can be trained to understand the depth of the characteristics in the image.

The Convolutional Neural network is designed with the intent of reducing the images into a form that is easier to process, without losing important features which are critical for getting a good prediction.

The architecture of the presented CNN network, as presented in Fig. 15.2, is composed of neurons that have weights and biases. Each neuron receives certain inputs and performs a scalar product. The whole network expresses a function with differentiable scores: the raw pixels in an image are taken at the input in order to predict the class scores at the output. The prediction is represented by the output class at the end of the chain, by a loss function, usually a Softmax function. Typically, a CNN is composed of an input and an output layer, between which we may have multiple hidden layers. The types of hidden layers of a CNN can be convolutional layers, pooling layers, fully connected layers or normalization layers [1].

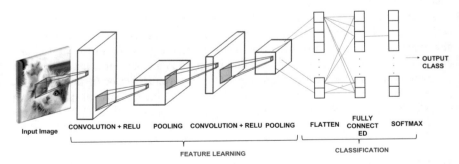

Fig. 15.2 The architecture of the ConvNet with alternative convolutional filters, pooling operations and fully connected layers are applied on the original data given as input

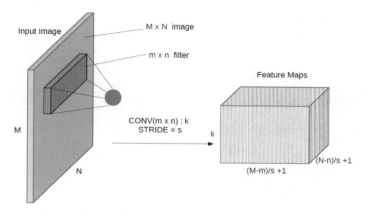

Fig. 15.3 The concept of convolutional filters with the corresponding parameters

- **Convolutional layers** are represented by a convolutional kernel with a stride, usually of one position, which is moved across the input data. The extraction of the most important features by the convolutional filters is represented in Fig. 15.3. In this way one can set the dimension of the kernel, $m \times n$, along with the total number of kernels k, taking into consideration the fact that a large number of kernels means better feature extraction capabilities, but this also means a large number of parameters that have to be trained increasing the time and memory necessary for the classification. The convolutional layers with stride one are used to reduce the spectral dimension.

- **Activation function** usually used for the ConvNet is ReLU. After each convolutional layer, an activation function called ReLU is used, having the capability of improving the network to exploit the local patterns by enforcing a local connectivity pattern between neurons of adjacent layers. The ReLU function, represented by $f(x) = max(0; x)$, is the activation that simply makes a threshold at zero. On each layer, a ReLU activation function is performed which accelerates the convergences of the gradient descent compared with other functions such as the sigmoid or tanh functions.

- **Downsampling layers**, known as POOLING layers, follow the convolutional layers. After each activation function of type ReLU is performed, a pooling layer follows. Each pooling layer is represented by a window of a certain dimension, for example a 2×2 window, slid along the data with the corresponding stride, for examples of 1 value as shown in Fig. 15.4. This window will retain only the most important information, in our case only the maximum values from the ones selected by the moving pooling window. The pooling layers are also known as downsampling layers capable of extracting the most important information

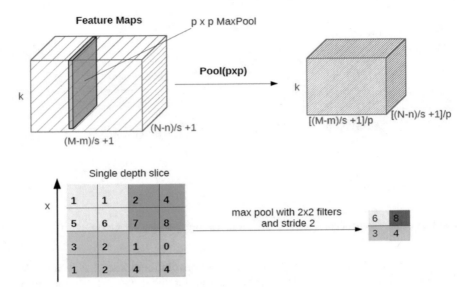

Fig. 15.4 The concept of pooling filters with the corresponding generated dimensions

from the input volume to reduce the computation cost and control the overfitting problems.

- **Fully connected layers** are usually placed at the end of the ConvNet architecture, when the output volume has a desirable form, typically a tensor of dimension $1 \times n$. This volume is then fed into a fully connected (FC) layer which uses a Soft-max function alongside the label information, representing the corresponding labels for each input sample. This section of the network will decide, based on the data received from the previous layers, what is the most likely correct label for the processed sample.

15.2   Applications and MATLAB Examples

15.2.1   Deep Learning MATLAB Toolbox

The Deep Network Designer app in Matlab allows one to build and edit deep learning networks in an interactive manner [2]. This app can import and edit networks, build new networks from scratch, even add new layers and create new connections. Also, it permits one to view and edit layer properties, combine networks and import custom layers.

To open the Deep Network Designer, one can type in the command line:

```
deepNetworkDesigner
```

15.2.2   Define, Train and Test a CNN Architecture

In this application we will use the Deep Network Designer to adapt a pretrained GoogLeNet network to classify a new collection of images [2]. This process is called transfer learning and is usually much faster and easier than training a new network. This is possible because one can apply learned features to a new task using a smaller number of training images.

Step 1. In the workspace, unzip the data and open the Deep Network Designer.

```
clear all
close all
clc

unzip('database.zip');
deepNetworkDesigner
```

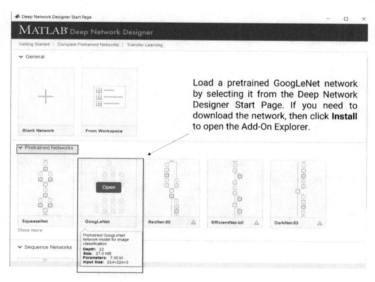

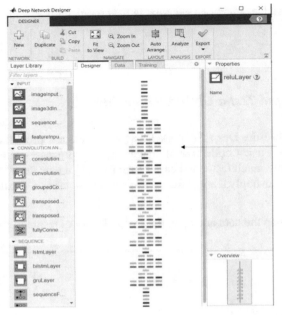

Deep Network Designer displays a zoomed-out view of the whole network. One can explore the network plot.

Step 2. Load Data Set.

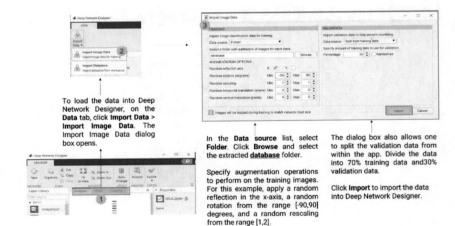

To load the data into Deep Network Designer, on the **Data** tab, click **Import Data > Import Image Data**. The Import Image Data dialog box opens.

In the **Data source** list, select **Folder**. Click **Browse** and select the extracted **database** folder.

Specify augmentation operations to perform on the training images. For this example, apply a random reflection in the x-axis, a random rotation from the range [-90,90] degrees, and a random rescaling from the range [1,2].

The dialog box also allows one to split the validation data from within the app. Divide the data into 70% training data and 30% validation data.

Click **Import** to import the data into Deep Network Designer.

• Deep Network Designer resizes the images during training to match the network input size. To view the network input size, in the Designer tab, click the `imageInputLayer`

Remark

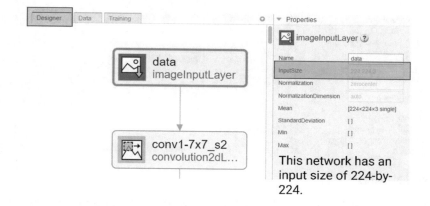

Step 3. Edit Network for Transfer Learning

To retrain a pretrained network to classify new images, one has to replace the last learnable layer and the final classification layer with new layers adapted to the new data set.

In GoogLeNet, these layers have the names 'loss3-classifier' and 'output'.

1. Replace the 'loss3-classifier' layer:

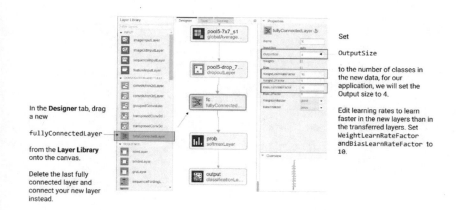

2.   Replace the 'output' layer:

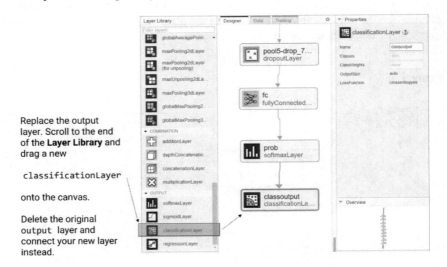

Replace the output
layer. Scroll to the end
of the **Layer Library** and
drag a new

classificationLayer

onto the canvas.

Delete the original
output layer and
connect your new layer
instead.

Check Network

Check your network by clicking **Analyze**. The network is ready for training if Deep
Learning Network Analyzer reports **zero errors**.

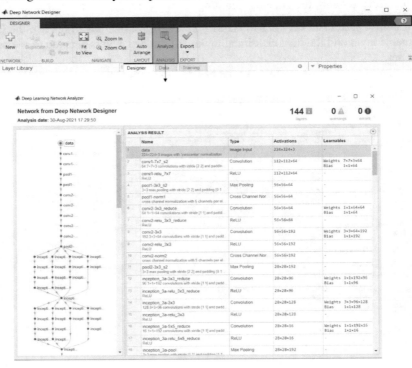

Step 4. Train the network

For our application, the **InitialLearnRate** to set to 0.0001, the **ValidationFrequency** is set to 3, and the **MaxEpochs** to 5. As there are 15 observations, set **MiniBatchSize** to 4 to divide the training data evenly and ensure the whole training set is used during each epoch. To train the network with the specified training options, click **Close** and then **Train**.

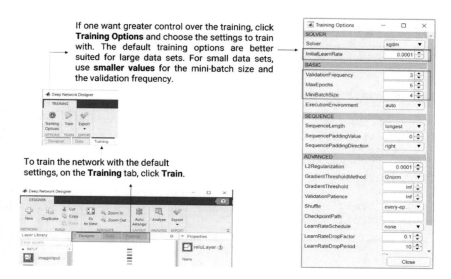

Deep Network Designer allows you to visualize and monitor the training progress. You can then edit the training options and retrain the network, if required.

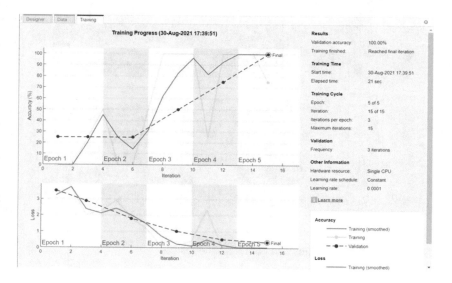

• To export the results from training, on the **Training** tab, select **Export**, then **Export Trained Network** and **Results. Deep Network Designer** exports the trained network as the variable **trained Network_1** and the training info as the variable **trainInfoStruct_1**

Remark

Step 5. Test and test the Network

```
% Select a new image to classify using the trained net-
work.
I = imread("dog13.jpg");

% Resize the test image to match the network input size.
I = imresize(I, [224 224]);

% Classify the test image using the trained network.
[YPred,probs] = classify(trainedNetwork_1,I);
imshow(I)
label = YPred;
title(string(label)  +  ",  "+ num2str(100*max(probs),3)
+"%");
```

Step 6. Test different images and comment the results.

15.3 Conclusions

This chapter presents a presentation of the convolutional neural network and an example of classifying different images based on GoogLeNet architecture in Matlab.

References

1. Li W, Prasad S, James F, Lour B (2012) Locality preserving dimensionality reduction and classification for hyperspectral image analysis. IEEE Trans Geosci Rem Sen 50(4):1185–1198
2. Matlab Help. Create Simple Deep Learning Network for Classification https://ch.mathwo rks.com/help/deeplearning/ug/create-simple-deep-learning-network-for-classification.html. Accessed 12 Sep 2021

Appendix A

Table of the primitive polynomials of degree k (k maximum $= 100$). The table presents only one primitive polynomial for each degree $k \leq 100$. In the table are the values of the exponents from the polynomial, e.g. $7\ 1\ 0 \rightarrow \times^7 + \times + 1$.

1	0						51	6	3	1	0		
2	1	0					52	3	0				
3	1	0					53	6	2	1	0		
4	1	0					54	6	5	4	3	2	0
5	2	0					55	6	2	1	0		
6	1	0					56	7	4	2	0		
7	1	0					57	5	3	2	0		
8	4	3	2	0			58	6	5	1	0		
9	4	0					59	6	5	4	3	1	0
10	3	0					60	1	0				
11	2	0					61	5	2	1	0		
12	6	4	1	0			62	5	3	0			
13	4	3	1	0			63	1	0				
14	5	3	1	0			64	4	3	1	0		
15	1	0					65	4	3	1	0		
16	5	3	2	0			66	8	6	5	3	2	0
17	3	0					67	5	2	1	0		
18	5	2	1	0			68	7	5	1	0		
19	5	2	1	0			69	6	5	2	0		
20	3	0					70	5	3	1	0		
21	2	0					71	5	3	1	0		
22	1	0					72	6	4	3	2	1	0

(continued)

M. Borda et al., *Randomness and Elements of Decision Theory Applied to Signals*, https://doi.org/10.1007/978-3-030-90314-5

(continued)

1	0						51	6	3	1	0		
23	5	0					73	4	3	2	0		
24	4	3	1	0			74	7	4	3	0		
25	3	0					75	6	3	1	0		
26	6	2	1	0			76	5	4	2	0		
27	5	2	1	0			77	6	5	2	0		
28	3	0					78	7	2	1	0		
29	2	0					79	4	3	2	0		
30	6	4	1	0			80	7	5	3	2	1	0
31	3	0					81	4	0				
32	7	5	3	2	1	0	82	8	7	6	4	1	0
33	6	4	1	0			83	7	4	2	0		
34	7	6	5	2	1	0	84	8	7	5	3	1	0
35	2	0					85	8	2	1	0		
36	6	5	4	2	1	0	86	6	5	2	0		
37	5	4	3	2	1	0	87	7	5	1	0		
38	6	5	1	0			88	8	5	4	3	1	0
39	4	0					89	6	5	3	0		
40	5	4	3	0			90	5	3	2	0		
41	3	0					91	7	6	5	3	2	0
42	5	4	3	2	1	0	92	6	5	2	0		
43	6	4	3	0			93	2	0				
44	6	5	2	0			94	6	5	1	0		
45	4	3	1	0			95	6	5	4	2	1	0
46	8	5	3	2	1	0	96	7	6	4	3	2	0
47	5	0					97	6	0				
48	7	5	4	2	1	0	98	7	4	3	2	1	0
49	6	5	4	0			99	7	5	4	0		
50	4	3	2	0			100	8	7	2	0		

Appendix B

Matlab code for the implementation from Chapter 9.

1. `probe_initial.m` used for calculating initial values

```
cdfFileName='HG-U133A';
fileNameCDF=strcat(cdfFileName,'.','cdf');
cdfStruct=affyread(fileNameCDF);
nGenes=cdfStruct.NumProbeSets;
nrc=cdfStruct.Cols;
%all the information of the experiment

%create the structure for the CEL files
%celFileName=sprintf('HG-U133B');
celFileName=sprintf('Tumores_A19T (A)HG-U133A');
fileNameCell=strcat(celFileName,'.','cel');
celStruct=affyread(fileNameCell);

GammaV = []; %gamma values for A19T file
vrel = []; %gamma <= 0,05, considered to be very reliable
rel = []; %gamma<=0,1 si >0,05, considered to be reliable
unrel = []; %gamma<=0,5 si >0.1, considered to be unreliable
vunrel = []; %gamma>0,05, considered to be very unreliable
GEU = [];
GEVU = [];

for i = 1:nGenes
   GeneName = cdfStruct.ProbeSets(i,1).Name;
   ProbeValues = probesetvalues(celStruct,cdfStruct,GeneName);
   a = modelo1(ProbeValues(:,7)',ProbeValues(:,14)');

   s(i,:).Gene=GeneName;
   s(i,:).PM=ProbeValues(:,7);
   s(i,:).MM=ProbeValues(:,14);


   Gamma = a(1);
   GammaV = [GammaV Gamma];
   if Gamma <= 0.05
```

```
      vrel = [vrel i];
   elseif 0.05 < Gamma && Gamma <= 0.1
      rel = [rel i];
   elseif 0.1 < Gamma && Gamma <= 0.5
      unrel = [unrel i];
   elseif 0.5 < Gamma
      vunrel = [vunrel i];
   end
end

GammaVUV = []; % very unreliable vector
for i = 1:nGenes
   if any(i==vunrel)
      GammaVUV = [GammaVUV GammaV(i)];
   end
end
GammaUV = []; %unreliable vector
for i = 1:nGenes
   if any(i==unrel)
      GammaUV = [GammaUV GammaV(i)];
   end
end

for i = 1:nGenes
   if any(i==vunrel)
      GeneName = cdfStruct.ProbeSets(i,1).Name;
      ProbeValues = probesetvalues(celStruct,cdfStruct,GeneName);
      clear ProbSet;
      ProbSet(1,:) = ProbeValues(:,7)';%PM
      ProbSet(2,:) = ProbeValues(:,14)';%MM

      %MAS 5.0
       k=0;
     for j = 1:length(ProbSet(1,:))
       if (ProbSet(1,j)-ProbSet(2,j)/(ProbSet(1,j)+ProbSet(2,j))
         >0.0015)
         k=k+1;
         x(k) = log2(ProbSet(1,j)) - log2(ProbSet(2,j));
       end
     end

      bg = tb(x);
    %Ideal Mismach
     for j = 1:length(ProbSet(1,:))
      if ProbSet(2,j)<ProbSet(1,j)
        IM_VU(j) = ProbSet(2,j);
      else
        if ProbSet(2,j)>ProbSet(1,j) && bg>0.03
         IM_VU(j) = ProbSet(1,j)/2^bg;
        else
         IM_VU(j) = ProbSet(1,j)/2^(0.03/(1+(0.03-bg)/10));
        end
      end
     end
```

```
   for j = 1:length(ProbSet(1,:))
     y(j) = log2(ProbSet(1,j) - IM_VU(j));
   end
  Exp = tb(y);%genetic expression computed by MAS 5.0
  GEVU = [GEVU, Exp];
 end
end

for i = 1:nGenes
  if any(i==unrel)
    GeneName = cdfStruct.ProbeSets(i,1).Name;
    ProbeValues = probesetvalues(celStruct,cdfStruct,GeneName);
    clear ProbSet;
    ProbSet(1,:) = ProbeValues(:,7)';%PM
    ProbSet(2,:) = ProbeValues(:,14)';%MM

    %MAS 5.0
    k=0;
  for j = 1:length(ProbSet(1,:))
    if (ProbSet(1,j)-ProbSet(2,j)/(ProbSet(1,j)+ProbSet(2,j))
       >0.0015)
      k=k+1;
      x(k) = log2(ProbSet(1,j)) - log2(ProbSet(2,j));
    end
  end

  bg = tb(x);
 %Ideal Mismach
  for j = 1:length(ProbSet(1,:))
   if ProbSet(2,j)<ProbSet(1,j)
    IM_U(j) = ProbSet(2,j);
   else
     if ProbSet(2,j)>ProbSet(1,j) && bg>0.03
      IM_U(j) = ProbSet(1,j)/2^bg;
     else
      IM_U(j) = ProbSet(1,j)/2^(0.03/(1+(0.03-bg)/10));
     end
   end
  end

  for j = 1:length(ProbSet(1,:))
    y(j) = log2(ProbSet(1,j) - IM_U(j));
   end
  ExpU = tb(y);%genetic expression computed by MAS 5.0
  GEU = [GEU, ExpU];
  end
end
save(celFileName,'nGenes','GammaV', 'vrel', 'rel', 'unrel',
'vunrel', 'a', 'GammaVUV', 'GammaUV', 'GEVU','GEU','s');
```

2.   probe_fastICA.m—calculating the values after fastICA was applied.

```
cdfFileName='HG-U133A';
fileNameCDF=strcat(cdfFileName,'.','cdf');
cdfStruct=affyread(fileNameCDF);
```

```matlab
nGenes=cdfStruct.NumProbeSets;
nrc=cdfStruct.Cols;

%create the structure for the CEL files
celFileName=sprintf('Tumores_A19T (A)HG-U133A');
fileNameCell=strcat(celFileName,'.','cel');
celStruct=affyread(fileNameCell);

GammaV = []; %gamma values for A19T file
vrel = []; %gamma <= 0,05, considered to be very reliable
rel = []; %gamma<=0,1 si >0,05, considered to be reliable
unrel = []; %gamma<=0,5 si >0.1, considered to be unreliable
vunrel = []; %gamma>0,05, considered to be very unreliable

GEU_Est = [];
GEU = [];
GEVU_Est = [];
GEVU = [];
GE_Est= [];

for i = 1:nGenes
   GeneName = cdfStruct.ProbeSets(i,1).Name;
   ProbeValues = probesetvalues(celStruct,cdfStruct,GeneName);
   a = modelo1(ProbeValues(:,7)',ProbeValues(:,14)');
   Gamma = a(1);
   GammaV = [GammaV Gamma];
   if Gamma <= 0.05
      vrel = [vrel i];
   elseif 0.05 < Gamma && Gamma <= 0.1
      rel = [rel i];
   elseif 0.1 < Gamma && Gamma <= 0.5
      unrel = [unrel i];
   elseif 0.5 < Gamma
      vunrel = [vunrel i];
   end
end

GammaVUV = []; %vectorul valorilor very unreliable
for i = 1:nGenes
   if any(i==vunrel)
      GammaVUV = [GammaVUV GammaV(i)];
   end
end
GammaUV = []; %vectorul valorilor unreliable
for i = 1:nGenes
   if any(i==unrel)
      GammaUV = [GammaUV GammaV(i)];
   end
end

GammaEVU    =    [];    %gamma    vector    for    very    unreli-
able values after FastICA is applied

for i = 1:nGenes
   if any(i==vunrel)
      GeneName = cdfStruct.ProbeSets(i,1).Name;
```

```
      ProbeValues = probesetvalues(celStruct,cdfStruct,GeneName);
      a = modelo1(ProbeValues(:,7)',ProbeValues(:,14)');
      Col_vector = a(2)*(ProbeValues(:,7)');
      Orth_vector = ProbeValues(:,14)' - Col_vector;
      X = [Col_vector; abs(Orth_vector)];
      Y = fastica(X);
      E = (-1)*[Y(1,:)./a(2); Y(1,:)+Y(2,:)];
      b = modelo1(E(1,:),E(2,:));
      GammaE = b(1);
      GammaEVU = [GammaEVU GammaE];
      clear ProbSet;
      ProbSet(1,:) = Y(1,:)./GammaE;
      ProbSet(2,:) = Y(1,:) + Y(2,:);

      s(i,:).Gene=GeneName;
      s(i,:).PM=ProbSet(1,:);
      s(i,:).MM=ProbSet(2,:);

      %MAS 5.0
       k=0;
     for j = 1:length(ProbSet(1,:))
        if (ProbSet(1,j)-ProbSet(2,j)/(ProbSet(1,j)+ProbSet(2,j))
           >0.0015)
          k=k+1;
          x(k) = log2(ProbSet(1,j)) - log2(ProbSet(2,j));
        end
     end

      bg = tb(x);
    %Ideal Mismach
     for j = 1:length(ProbSet(1,:))
      if ProbSet(2,j)<ProbSet(1,j)
       IM(j) = ProbSet(2,j);
      else
         if ProbSet(2,j)>ProbSet(1,j) && bg>0.03
          IM(j) = ProbSet(1,j)/2^bg;
         else
          IM(j) = ProbSet(1,j)/2^(0.03/(1+(0.03-bg)/10));
         end
      end
     end

     for j = 1:length(ProbSet(1,:))
      y(j) = log2(ProbSet(1,j) - IM(j));
     end
     Exp = tb(y);%genetic expression computed with MAS 5.0
     GEVU_Est= [GEVU_Est Exp];
    end
end

GammaEU = []; %gamma unreliable vector after FastICA was applied
for i = 1:nGenes
   if any(i==unrel)
     GeneName = cdfStruct.ProbeSets(i,1).Name;
     ProbeValues = probesetvalues(celStruct,cdfStruct,GeneName);
```

```
    a = modelo1(ProbeValues(:,7)',ProbeValues(:,14)');
    Col_vector = a(2)*(ProbeValues(:,7)');
    Orth_vector = ProbeValues(:,14)' - Col_vector;
    X = [Col_vector; abs(Orth_vector)];
    Y = fastica(X, 'approach', 'symm', 'g', 'pow3');
    E = (-1)*[Y(1,:)./a(2); Y(1,:)+Y(2,:)];
    b = modelo1(E(1,:),E(2,:));
    GammaE = b(1);
    GammaEU = [GammaEU GammaE];
    clear ProbSet;
    ProbSet(1,:) = Y(1,:)./GammaE;
    ProbSet(2,:) = Y(1,:) + Y(2,:);

    s(i,:).Gene=GeneName;
    s(i,:).PM=ProbSet(1,:);
    s(i,:).MM=ProbSet(2,:);

    %MAS 5.0
     k=0;
  for j = 1:length(ProbSet(1,:))
    if (ProbSet(1,j)-ProbSet(2,j)/(ProbSet(1,j)+ProbSet(2,j))
       >0.0015)
       k=k+1;
       x(k) = log2(ProbSet(1,j)) - log2(ProbSet(2,j));
    end
  end

  bg = tb(x);
 %Ideal Mismach
  for j = 1:length(ProbSet(1,:))
   if ProbSet(2,j)<ProbSet(1,j)
    IM(j) = ProbSet(2,j);
   else
     if ProbSet(2,j)>ProbSet(1,j) && bg>0.03
      IM(j) = ProbSet(1,j)/2^bg;
     else
      IM(j) = ProbSet(1,j)/2^(0.03/(1+(0.03-bg)/10));
     end
   end
  end

   for j = 1:length(ProbSet(1,:))
    y(j) = log2(ProbSet(1,j) - IM(j));
   end
  Exp = tb(y);%genetic expression computed by MAS 5.0
  GEU_Est= [GEU_Est Exp];
  end
end

file=strcat(celFileName,'_Fastica');

save(file,'nGenes','GammaV','vrel','rel','unrel','vunrel','a',
'GammaVUV',       'GammaUV',       'GammaEVU',       'GammaEU',
'GEVU_Est','GEU_Est','s');
```

Appendix C

Introduction to Matlab

MATLAB® = High Performance Language for Computer Aided Design.

MATLAB is both a ***programming language*** and a ***development system*** that integrates computation, visualization and programming in an easy-to-use environment, with problems and their solutions expressed in an accessible mathematical language.

Areas of usage:

- Mathematics and numerical calculation
- Algorithm development
- Modelling, simulation and prototype testing
- Data analysis and visualization
- Engineering and applied science graphics
- Application development, including GUI

MATLAB = an interactive system based on arrays and matrices, which enables numerical computation problems to be solved, especially those requiring the processing of vectors or matrices.

- MATLAB = an interactive system based on arrays and matrices, which enables numerical computation problems to be solved, especially those requiring the processing of vectors or matrices.
- The name MATLAB comes from ***Mat**rix **lab**oratory*
- The manufacturer is **The MathWorks, Inc., SUA**
- MATLAB has evolved:

    - In academia area where is the standard package for introductory and advanced math courses, engineering, and science.
    - in industry, where it is used for high-throughput research, development and manufacturing

M. Borda et al., *Randomness and Elements of Decision Theory Applied to Signals*, https://doi.org/10.1007/978-3-030-90314-5

- MATLAB enables the development of a family of applications in the form of toolboxes. These toolboxes enable the learning and application of specialized technologies in various fields. Toolboxes are available for areas such as: numerical signal processing, automatic driving systems, neural networks, fuzzy logic, wavelet, simulation (SIMULINK), identification, etc.

The **MATLAB system** consists of five main:

- **MATLAB language**
- **The MATLAB environment**
- **Handle Graphics®**
- **MATLAB's library of mathematical functions**
- **MATLAB Application Program Interface (API)**.

MATLAB Language: represents a high-level matrix/table language with jump control statements, functions, data structures, I/O, and object-oriented programming properties. The programming facilities are organized into 6 directories:

Ops	Operators and special characters
Lang	Programming language constructs
Strfun	Character strings
Iofun	File input/output
Timefun	Time and dates
Datatypes	Data types and structures

MATLAB Environment: A set of facilities that allow you to manipulate variables in the workspace, import and export data, develop, manipulate, edit and debug MATLAB files (.m) and MATLAB applications. These facilities are organized in the directory:

General	General purpose commands

Handle Graphics®: Represents the MATLAB graphical system. It includes high-level commands for two- and three-dimensional data visualization, image processing, animation, and graph presentations. It also allows the use of low-level commands to create GUI graphical interfaces. The graphical functions are organised in 5 directories:

graph2d	Two-dimensional graphs
Graph3d	Three-dimensional graphs
Specgraph	Specialized graphs
Graphics	Handle graphics
Uitools	Graphical user interface tools

MATLAB Mathematical Function Library: represents a complex collection of computational algorithms ranging from elementary functions (sine, cosine, etc.) to sophisticated functions (matrix inversion, eigenvalues, Bessel functions, FFT, etc.). The mathematical functions are organised in 8 directories:

elmat	Elementary matrices and matrix manipulation
Elfun	Elementary math functions
Specfun	Specialized math functions
Matfun	Matrix functions—numerical linear algebra
Datafun	Data analysis and fourier transforms
Polyfun	Interpolation and polynomials
Funfun	Function functions and ODE solvers
Sparfun	Sparse matrices

MATLAB's Application Program Interface (API) is a library that allows you to write programs in C or Fortran that interact with MATLAB. It includes facilities for calling MATLAB routines, calling MATLAB as a computing machine, writing and reading.MAT files.

SIMULINK Package

- SIMULINK® is a software package attached to MATLAB and is an interactive system for simulating the dynamics of nonlinear (and of course linear) systems. It is designed as a graphical interface that allows you to create a model by "drawing" the block diagram of the system and then simulating the dynamics of the system.
- SIMULINK can work with linear, non-linear, continuous, discrete, multivariable, etc. systems.
- SIMULINK makes use of so-called Blocksets which are in fact additional libraries containing specialized applications in areas such as: communications, signal processing, etc.
- Real-time Workshop® is a very important program that allows the generation of C code for the block schemas created in SIMULINK and therefore allows the running of a wide variety of real-time applications.

MATLAB Toolboxes: Toolboxes are a family of applications that allow learning and application of specialized technologies in various fields. These toolboxes are collections of MATLAB functions (M-files) that extend the MATLAB environment for solving particular classes of problems. Some of the most commonly used applications are shown in the following figure.

Fundamental Operators

MATLAB works with mathematical expressions like other programming languages, but unlike most of these languages, these expressions involve working extensively with matrices.

Expressions are composed using the following types:

- Variables
- Numbers
- Operators
- Functions

Variables

- MATLAB does not require you to declare the size of variables because when a new variable name is encountered it automatically generates that variable and allocates the required memory space.
- A variable name is a letter followed by any number of letters, digits, or symbols. From this "any number" the first 31 characters are stopped.
- MATLAB is *case sensitive*—it distinguishes between lowercase and uppercase letters.

```
» a = 30
```
creates a 1 × 1 array *a* and stores its value 30 in a single location corresponding to the single element of the matrix

Example

– *Numbers*

MATLAB uses decimal notation, with optional decimal point and + or − sign. Scientific notation with the letter e is also used to specify a power of 10. Representation of imaginary numbers is done with the letter i or j as a suffix.

```
3              −99            0.0001
9.6397238      1.60210e-20    6.02252e23
1i             −3.14159j      3e5i
```

Example

- All numbers are stored internally using the *long* format specified by the IEEE floating point standard (precision of 16 significant decimal places in the range 10^{-308} to 10^{+308}).

– *Operators*

Expressions use common arithmetic operators:

+	Addition
−	Subtraction
*	Multiplication
/	Division
\	Left division
^	Raising to a power
'	Transposed complex conjugate
()	Operator specifying the order of evaluation

– *Functions*

MATLAB provides a large number of standard elementary mathematical functions (abs, sqrt, exp, sin …).

There are also advanced mathematical functions (Bessel functions, gamma, etc.), many of which accept complex arguments.

To view the elementary functions you can type:

```
» help elfun
```

To see the list of advanced functions you can type:

```
» help specfun
» help elmat
```

- Some of the functions (such as sqrt, sin) are built-in, i.e. they are part of the MATLAB core, they are highly efficient, but the construction details are not accessible to the user.
- Other functions are implemented as MATLAB files (M-files) and can even be modified.
- Some functions provide the values of universal constants:

pi	3.14159265
I	Imaginary unit, -1
J	Same as I
Eps	Floating-point relative precision, 2^{-52}
Realmin	Smallest floating-point number, 2^{-1022}
Realmax	Largest floating-point number, 2^{1023}
Inf	Infinity
NaN	Not-a-number

- The function names are not reserved and can therefore be overwritten.

eps = 1.e-6
The original function is reconstructed by the command:
» clear eps

Example

– *Expressions*

Examples of expressions and the corresponding results of the evaluation of these expressions:

```
» rho = (1+sqrt(5))/2
rho =
    1.6180
» a = abs(3+4i)
a =
    5
» z = sqrt(besselk(4/3,rho-i))
z =
   0.3730+ 0.3214i
» huge = exp(log(realmax))
huge =
  1.7977e+308
» toobig = pi*huge
toobig =
   Inf
```

Data Format and Saving Options

Help on-Line

To run MATLAB on a PC, simply double-click on the MATLAB icon with the mouse. If the operating system is not Windows (it is UNIX) you must type matlab after the operating system prompt.

- The MATLAB language is much simpler to learn if one leaves out the arid inspection of lists of variables, functions and operators and instead uses the help, helpdesk, demo commands typed directly from the MATLAB prompt.
- To find out all useful information about a command or function, type help followed by the name of that command or function.
- The MATLAB package also has complete usage information in the form of a.pdf documentation.
- In particular cases, the INTERNET can be used and there is a link to the manufacturer's website.
- Other useful commands for finding information are: helpwin, lookfor, help help.

Suggestive examples of how to use the help command:

```
» help sin
 SIN    Sine.
   SIN(X) is the sine of the elements of X.

 Overloaded methods
    help sym/sin.m
» help exp

 EXP    Exponential.
   EXP(X) is the exponential of the elements of X,
 e to the X.
   For complex Z=X+i*Y, EXP(Z) =
 EXP(X)*(COS(Y)+i*SIN(Y)).

   See also LOG, LOG10, EXPM, EXPINT.

 Overloaded methods
    help sym/exp.m
    help demtseries/exp.m

 » help plot

 PLOT   Linear plot.
```

PLOT(X,Y) plots vector Y versus vector X. If X or Y
is a matrix, then the vector is plotted versus the rows
or columns of the matrix, whichever line up. If X is a
scalar and Y is a vector, length(Y) disconnected points
are plotted.

PLOT(Y) plots the columns of Y versus their index.
If   Y   is   complex,   PLOT(Y)   is   equivalent   to

PLOT(real(Y),imag(Y)). In all other uses of PLOT, the
imaginary part is ignored.

Various line types, plot symbols and colors may be
obtained with PLOT(X,Y,S) where S is a character string
made from one element from any or all the following 3
colunms:

```
y    yellow        .    point        -    solid
m    magenta       o    circle       :    dotted
c    cyan          x    x-mark       -.   dashdot
r    red           +    plus         --   dashed
g    green         *    star
b    blue          s    square
w    white         d    diamond
```

```
k    black          v    triangle (down)
                    ^     triangle (up)
                    <    triangle (left)
                    >    triangle (right)
                    p    pentagram
                    h    hexagram
```

For example, PLOT(X,Y,'c+:') plots a cyan dotted line with a plus at each data point; PLOT(X,Y,'bd') plots blue diamond at each data point but does not draw any line.

PLOT(X1,Y1,S1,X2,Y2,S2,X3,Y3,S3,...) combines the plots defined by the (X,Y,S) triples, where the X's and Y's are vectors or matrices and the S's are strings.

For example, PLOT(X,Y,'y-',X,Y,'go') plots the data twice, with a solid yellow line interpolating green circles at the data points.

The PLOT command, if no color is specified, makes automatic use of the colors specified by the axes ColorOrder property. The default ColorOrder is listed in the table above for color systems where the default is yellow for one line, and for multiple lines, to cycle through the first six colors in the table. For monochrome systems, PLOT cycles over the axes LineStyleOrder property.

PLOT returns a column vector of handles to LINE objects, one handle per line.

The X,Y pairs, or X,Y,S triples, can be followed by parameter/value pairs to specify additional properties of the lines.

See also SEMILOGX, SEMILOGY, LOGLOG, GRID, CLF, CLC, TITLE, XLABEL, YLABEL, AXIS, AXES, HOLD, COLORDEF, LEGEND, and SUBPLOT.

Data Format

MATLAB displays numbers to 5 decimal (default setting). This setting can be changed using the command format:

FORMAT Set output format.
All computations in MATLAB are done in double precision.
 FORMAT may be used to switch between different output
 display formats as follows:
 FORMAT   Default. Same as SHORT.
 FORMAT SHORT   Scaled fixed point format with 5 digits.
 FORMAT LONG   Scaled fixed point format with 15 digits.
 FORMAT SHORT E Floating point format with 5 digits.
 FORMAT LONG E Floating point format with 15 digits.
 FORMAT SHORT G Best of fixed or floating point format with
5 digits.
 FORMAT LONG G Best of fixed or floating point format with
15 digits.
 FORMAT HEX   Hexadecimal format.
 FORMAT +   The symbols +, - and blank are printed
                for positive, negative and zero elements.
                Imaginary parts are ignored.
 FORMAT BANK   Fixed format for dollars and cents.
 FORMAT RAT   Approximation by ratio of small integers.
 Spacing:
 FORMAT COMPACT Suppress extra line-feeds.
 FORMAT LOOSE   Puts the extra line-feeds back in.

Example

```
» c=1.333456789233
c =
1.3335
» format long
» c
c =
1.33345678923300
» format short e
» c
c =
1.3335e+000
» format long e
» c
c =
1.333456789233000e+000
» format
» c
c =
1.3335
```

Saving Options

- The save command in MATLAB can be used to save the current variables at the end of a work session.
- This command will save all current user-generated variables to a file called *matlab.mat*. If desired you can give a name to the data file where the variables are saved.

Example

```
» save date c determ A
```
saves data c determ and A to a *data.mat* file
• The load command is used to return variables in a subsequent
work session. Example:
```
» load date
```
• If you want to find out the current variables you can use the
who,whos commands:
```
» who
Your variables are:
A            c            determ
» whos
Name            Size            Bytes   Class
A        2 x 2      32    double array
c            1 x 1       8     double array
determ   1 x 1       8     double array
Grand total is 6 elements using 48 bytes
```
• The clear command can be used to clear all current variables
from working memory

File Creations

Because it is much more convenient and useful than entering line-by-line commands
at the MATLAB prompt, we work with text files that contain these program lines
with the necessary commands.

These files contain code in the MATLAB language and are called.m files (or M-
files). The files are created using a text editor and then used as a regular MATLAB
command.

There are two types of .m files:

• Script files, which do not support input arguments and do not return output
arguments. These files operate with workspace data.
• Routines (functions), which accept input arguments and return output arguments.
The variables used are local (internal) variables of the function.

To see the contents of a MATLAB file, for example decision.m, use the
command:
```
» type decision.
```

Script Files

When a script file is used, MATLAB executes the commands found in that file. Script
files can work with workspace data or create new data with which they operate. The
scripts do not provide output arguments, and the variables created remain in the
workspace, to be used in subsequent calculations.

Script files can provide graphical outputs using functions such as plot and bar.

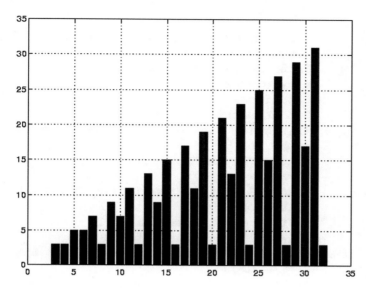

Fig. C.1 Example of a graph in Matlab

Example

Example script file: `magicrank.m`, with the following MATLAB commands:

```
% Investigate the rank of magic squares
r = zeros(1,32);
for n = 3:32
r(n) = rank(magic(n));
end
r
bar(r)
```

When typing the name of the script file (without the.m extension):

```
» magicrank
```

The MATLAB executes the commands, calculates the rank of some matrices (magic matrices), and draws the graph with the calculation results. After the execution of the file is finished, the variables n and r remain in the workspace. The resulting graph is shown in Fig. C.1.

Functions (Routines)

These files support input arguments and provide output arguments. The name of the MATLAB file (M-file) and that of the respective function (subroutine) must be identical. Functions (subroutines) work with their own variables separate from the usual MATLAB workspace.

Example

Example: rank function. The M-file rank.m file is available in the directory toolbox/matlab/matfun

The file can be viewed with the code:

```
» type rank
function r = rank(A,tol)
% RANK Matrix rank.
%   RANK(A) provides an estimate of the
number of
% linearly independent rows or columns of
a matrix A
% RANK(A,tol) is the number of singular
values of A
%   that are larger than tol.
% RANK(A) uses the default
%   tol = max(size(A)) * norm(A) * eps.
s = svd(A);
if nargin==1
tol = max(size(A)) * max(s) * eps;
end
r = sum(s > tol);
```

The first line of an M-file function begins with the function keyword. This line gives the function name, the order and the number of arguments.

The following lines (starting with the % character) are comment lines, which are actually the lines displayed when the command is called:

```
» help rank
```

The rest of the lines are executable. The variables s, like r, A, tol are local variables of the function and are separated from the variables in the workspace.

The rank function can be used in different ways:

```
» rank(A)
» r = rank(A)
» r = rank(A,1.e-6)
```

Global Variables

If several such subroutines are to use a certain common variable, declare that variable as global using the global command in all those functions.

Example: the file falling.m:
```
function h = falling(t)
global GRAVITY
h = 1/2*GRAVITY*t.^2;
```
the lines are then entered interactively:
```
» global GRAVITY
» GRAVITY = 32;
» y = falling((0:.1:5)');
```

Example

Example

Eval Function

The `eval` function works with text variables to implement a powerful macro text feature.

Expression:

```
eval(s)
```

uses the MATLAB interpreter to evaluate the expression or the execution of the statement in the string s.

Vectorization

In order to obtain a high computational speed, the so-called vectorization of algorithms in MATLAB files is very important. Where other languages use for or DO loops, MATLAB can use matrix or vector operations.

A simple example is the following:

```
x = 0;
for k = 1:1001
  y(k) = log10(x);
  x = x + .01;
```

The vectorized version of the same program is

```
x = 0:.01:10;
y = log10(x);
```

Matrices, Vectors and Polynomials

In order to work easily and well with the MATLAB language, you must first learn how to handle matrices. In MATLAB, an array is a rectangular array of numbers. Scalars, for example, are 1×1 matrices, and single-row or column matrices are actually vectors.

A famous example of the matrix appears in the Renaissance engraving Melancholia by the great amateur artist and mathematician Albrecht Dürer. The engraving is loaded with mathematical symbolism and a careful observation of it can be distinguished in the upper right corner of a matrix.

This matrix is known as the magic square and in Dürer's time it was considered to have magical properties.

Introduction of Matrices

Arrays can be written in several ways.

- Write a list of matrix elements.
- Upload data from external data files.
- Generating arrays using built-in functions.
- Creating arrays in M-files.

We will introduce Dürer's matrix first as a list of elements.
There are a few simple conventions to follow:

- The elements of a line are separated by commas or spaces.
- The end of a line is marked with a semicolon.
- The list of elements that make up the matrix is delimited in square brackets: [ ]

For Dürer's matrix we type:

```
» A = [16 3 2 13; 5 10 11 8; 9 6 7 12; 4 15 14 1]
```

In the command line we have the following result:

```
A =
   16    3    2   13
    5   10   11    8
    9    6    7   12
    4   15   14    1
```

A certain element of the matrix, for example the element in line i column j is denoted A (i, j).
Therefore another (less fast) way to calculate the amount on the fourth column for example is the following:

```
» A(1,4) + A(2,4) + A(3,4) + A(4,4)
ans =
   34
```

If we specify an element that does not exist in the array, we receive an error message:

```
» t = A(4,5)
 Index exceeds matrix dimensions.
```

Operator

Operator: It is very important. For example, the expression

```
»1:10
```

is a line vector

```
ans =
1   2   3   4   5   6   7   8   9   10
```

Other examples:

```
» 100:-7:50
ans =
    100    93    86    79    72    65    58    51
```

```
» 0:pi/4:pi
ans =
    0    0.7854    1.5708    2.3562    3.1416
```

Expression

```
A(1:k,j)
```

It refers to the first k elements of column j of A.

If it is used in parentheses the operator: then it means that we refer to all the elements of a line or column.

```
» sum(A(:,3))
```

calculates the sum of the elements in the third column of A:

```
ans =
   34
```

Magic function

Matlab has a built-in function that creates magic squares of any size:

```
» B = magic(4)
B =
   16    2    3   13
    5   11   10    8
    9    7    6   12
    4   14   15    1
```

This matrix is almost identical to Dürer's matrix, the only difference being that the two middle columns are interchanged. The following MATLAB command can be used to obtain the Dürer matrix from B:

```
» A = B(:,[1 3 2 4])
A =
   16    3    2   13
    5   10   11    8
    9    6    7   12
    4   15   14    1
```

Polynomials

- Polynomials are described in MATLAB by line vectors whose elements are in fact the coefficients of polynomials in descending order of powers.

Example: the polynomial $p(x) = x^3 + 5x + 6$ is represented in MATLAB as follows:

```
p = [1 0 5 6]
```

- A polynomial can be evaluated for a value of x using the `polyval` function:

```
» polyval(p,1)
ans=
      12
```

In the example above the polynomial p at point x = 1 is evaluated.

- You can easily find the roots of the polynomial using the `roots` function:

```
» r=roots(p)
r =
   0.5000 + 2.3979i
   0.5000 - 2.3979i
  -1.0000
```

There are many other functions and commands that deal with operations on polynomials, functions that will be addressed in a special chapter. Among them we mention the command that allows the multiplication of two polynomials, namely `conv`:

```
» p1=[1 3 5]
p1 =
     1    3    5

» p2=[2 0 1 0 5]
p2 =
     2    0    1    0    5

» p3=conv(p1,p2)
p3 =
          2    6   11    3   10   15   25
```

Basic Operations with Matrices and Functions

The MATLAB operates with matrices as easily as it works with scalars. For the addition of two matrices, for example, the + sign is simply used as in a regular addition. Of course, the matrices must have the same dimensions in order to be added.

Example:

```
» A=[2 3;15 -3]
A =
     2    3
    15   -3

» B=[11 -21; 12 4]
B =
    11   -21
```

```
   12     4
» C=A+B
C =
      13    -18
      27   1
```

The operator * is used to multiply two matrices, which is also valid for scaling operations. Example:» D = A*B.

```
D =
    58 -30
   129     -327
```

If the dimensions of the multiplying matrices are not appropriate, then an error message will be provided:

```
» E=[1 23; -12 2;1 2]
E =
     1    23
   -12     2
     1     2
» F=A*E
??? Error using ==> *
Inner matrix dimensions must agree.
```

To "troubleshoot" the program in case of such errors you can use the size command which gives us information about the dimensions of the respective matrices and allows the correction of errors:

```
» size(A)
ans =
     2    2
» size(E)
ans =
     3     2
```

The MATLAB includes many other array-operated functions that will be described and used extensively in the following chapters. We mention here a few: det, inv, rank, eig etc.

An interesting feature of MATLAB is that it works with matrices with logical and relational operators in a similar way to these operations performed with scalars.

For example, for scalar operation

```
» r=17>55
r =
    0
```

The MATLAB responds with r = 0, i.e. false. If, for example, we want to compare each element of the matrix A with the corresponding element in the matrix B, we proceed similarly:

```
» L=A<=B
```

```
L =
   1    0
   0    1
```

The logical operators like & for AND, | for OR, ~ for NOT, will return the value 1 for TRUE and 0 for FALSE. For example:

```
» A&B
ans =
   1    1
   1    1
» ~A
ans =
      0    0
      0    0
```

Printed in the United States
by Baker & Taylor Publisher Services